Handbook of IoT and Blockchain

Internet of Everything (IoE)

Series Editors: Vijender Kumar Solanki, Raghvendra Kumar, and Le Hoang Song

IOT
Security and Privacy Paradigm
Edited by Souvik Pal, Vicente Garcia Diaz and Dac-Nhuong Le

Smart Innovation of Web of Things
Edited by Vijender Kumar Solanki, Raghvendra Kumar and Le Hoang Son

Big Data, IoT, and Machine Learning
Tools and Applications
Rashmi Agrawal, Marcin Paprzycki, and Neha Gupta

Internet of Everything and Big Data
Major Challenges in Smart Cities
Edited by Salah-ddine Krit, Mohamed Elhoseny, Valentina Emilia Balas, Rachid Benlamri, and Marius M. Balas

Bitcoin and Blockchain
History and Current Applications
Edited by Sandeep Kumar Panda, Ahmed A. Elngar, Valentina Emilia Balas, and Mohammed Kayed

Privacy Vulnerabilities and Data Security Challenges in the IoT
Edited by Shivani Agarwal, Sandhya Makkar, and Tran Duc Tan

Handbook of IoT and Blockchain
Methods, Solutions, and Recent Advancements
Edited by Brojo Kishore Mishra, Sanjay Kumar Kuanar, Sheng-Lung Peng, and Daniel D. Dasig, Jr.

Blockchain Technology
Fundamentals, Applications, and Case Studies
Edited by E Golden Julie, J. Jesu Vedha Nayahi, and Noor Zaman Jhanjhi

For more information about this series, please visit: https://www.crcpress.com/Internet-of-Everything-IoE-Security-and-Privacy-Paradigm/book-series/CRCIOESPP

Handbook of IoT and Blockchain

Methods, Solutions, and Recent Advancements

Edited by
Brojo Kishore Mishra,
Sanjay Kumar Kuanar,
Sheng-Lung Peng,
and Daniel D. Dasig, Jr.

CRC Press is an imprint of the
Taylor & Francis Group, an informa business

First edition published 2020
by CRC Press
6000 Broken Sound Parkway NW, Suite 300,
Boca Raton, FL 33487-2742

and by CRC Press
2 Park Square, Milton Park, Abingdon, Oxon OX14 4RN

Library of Congress Cataloging-in-Publication Data
Names: Mishra, Brojo Kishore, 1979– editor. | Kuanar, Sanjay Kumar, editor. |
Peng, Sheng-Lung, editor. | Dasig, Daniel D., Jr., editor.
Title: Handbook of IoT and blockchain : methods, solutions,
and recent advancements / edited by Brojo Kishore Mishra,
Sanjay Kumar Kuanar, Sheng-Lung Peng, and Daniel D. Dasig, Jr.
Description: First edition. | Boca Raton, FL : CRC Press, 2021. |
Series: Internet of everything (IoE). Security and privacy paradigm |
Includes bibliographical references and index.
Identifiers: LCCN 2020022756 (print) | LCCN 2020022757 (ebook) |
ISBN 9780367422455 (hardback) | ISBN 9780367854744 (ebook)
Subjects: LCSH: Blockchains (Databases) | Internet of things.
Classification: LCC QA76.9.B56 H345 2021 (print) |
LCC QA76.9.B56 (ebook) | DDC 005.74–dc23
LC record available at https://lccn.loc.gov/2020022756
LC ebook record available at https://lccn.loc.gov/2020022757

ISBN: 978-0-367-42245-5 (hbk)
ISBN: 978-0-367-85474-4 (ebk)

Contents

Preface

Blockchain technology is receiving growing attention from various organizations and researchers as it provides magical solutions to the problems associated with classical centralized architecture. Blockchain, whether public or private, is a distributed ledger with the capability of maintaining the integrity of transactions by decentralizing the ledger among participating users.

On the other hand, the Internet of Things (IoT) represents a revolution of the internet which can connect nearly all environment devices over the internet to share their data to create novel services and applications for improving our quality of life. Although the centralized IoT system provides countless benefits, it raises several challenges. Resolving these challenges can be done by integrating IoT with blockchain technology.

This book will provide an overview of IoT and Blockchain concepts, IoT Solutions with Blockchain in the current business scenario and many more.

Editors

Brojo Kishore Mishra is currently working as professor in Computer Science and Engineering department at the GIET University, Gunupur-765022, India. He received his PhD degree in Computer Science from Berhampur University in 2012. He has published more than 30 research papers in national and international conference proceedings, 25 research papers in peer-reviewed journals, 22 book chapters, authored two books and edited four books. His research interests include Data mining, Machine learning, Soft computing, and Security

Sanjay K. Kuanar is currently serving as professor and head of Computer Science and Engineering Department of GIET University. He earned his Ph.D. degree from Jadavpur University, Kolkata in November, 2015, obtained his ME (Computer Engineering) degree from Jadavpur University in 2007 and B.Tech Degree in Computer Science and Engineering from Utkal University in 1998. His research interests include computer vision, pattern recognition, machine learning and multimedia computing. His primary research is focused on exploring different machine-learning techniques in solving several challenging problems in summarizing and segmenting videos. He has (co)authored several publications in refereed SCI journals including IEEE Transactions on Multimedia, IEEE Transactions on Cybernetics, Elsevier's JVCIR and international conferences such as ICPR, ICAPR etc. His papers have received many citations with h-index and i-10 index of 4 each. He has reviewed several international journals such as IEEE Transactions on Multimedia (TMM), IEEE Transactions on Cybernetics and Elsevier's JVCIR. His total teaching experience is 20 years. Dr. Kuanar is a self-motivated and self-driven leader and has taken several initiatives for the development of the department.

Sheng-Lung Peng is a professor of the Department of Creative Technologies and Productive Design at National Taipei University of Business, Taiwan. He received a BS degree in Mathematics from National Tsing Hua University, and MS and PhD degrees in Computer Science and Information Engineering from the National Chung Cheng University and National Tsing Hua University, Taiwan, respectively. His research interests are in designing and analyzing algorithms for Bioinformatics, Combinatorics, Data Mining, and Networks.

Dr. Peng has edited several special issues for journals, such as Soft Computing, Journal of Internet Technology, Journal of Computers and MDPI Algorithms. He is also a reviewer for many journals such as IEEE Access and Transactions on Emerging Topics in Computing, IEEE/ACM Transactions on Networking, Theoretical Computer Science, Journal of Computer and System Sciences, Journal of Combinatorial Optimization, Journal of Modeling in Management, Soft Computing, Information Processing Letters, Discrete Mathematics, Discrete Applied Mathematics, Discussiones Mathematicae Graph Theory, and so on. He has published more than 100 international conferences and journal papers.

Dr. Peng is now the dean of the Library and Information Services Office of NDHU, an honorary Professor of Beijing Information Science and Technology University of China, and a visiting professor of Ningxia Institute of Science and Technology of China. He is now the regional director of the ACM-ICPC Contest Council for Taiwan, a director of Institute of Information and Computing Machinery (IICM), of Information Service Association of Chinese Colleges and of Taiwan Association of Cloud Computing (TACC). He is also a supervisor of Chinese Information Literacy Association, of Association of Algorithms and Computation Theory (AACT), and of Interlibrary Cooperation Association in Taiwan. He has been serving as a secretary general of TACC from 2011 to 2015,of AACT from 2013 to 2016, and of IICM from 2015 to 2018. He was a convener of the East Region of Service Science Society of Taiwan from 2014 to 2016.

Daniel D. Dasig, Jr. received his BSc in Computer Engineering, MSc in Engineering in Computer Engineering, and PhD in Electrical and Computer Engineering degrees in 2008, 2015 and 2017 respectively. He has been a professorial lecturer in electrical engineering, environmental engineering, information technology, computer science, and education in the Philippines and abroad. He is an associate professor 6 of the Graduate Studies of College of Science and Computer Studies of De La Salle University Dasmarinas since January 2018, and a professorial lecturer at the College of Continuing and Advanced Professional Studies of the University of Makati since 2014 where he also served as research coordinator in the College of Computer Science. In 2014, he served as chair of Computer Engineering Department of Jose Rizal University. He also works as part of the change management, and software development teams of TELUS, a Canadian telecommunication company, since 2011. He is a Lean Six Sigma Certified, ITIL V3 Certified and Certified Project Management Expert. Dr. Dasig has authored and co-authored more than 100 academic and technical papers extending over a wide spectrum of research areas such as in engineering, information technology, education, computer science and social sciences. He has authored and co-authored books for higher education and senior high school programs. He has organized and co-organized local and international conferences and was a recipient of Best Paper and Best Paper Presenter awards and fellow. He has executive roles and active membership in more than 40 professional, research and policy-making organizations such as ICANN-NCSG, Internet Society, ICpEP, and PSITE. He has received numerous global and international awards including the World Class Filipino for Engineering and Information Technology. He is an IEEE member of the Republic of the Philippines Section.

1 Blockchain-Enabled Security and Privacy Schemes in IoT Technologies

Siddhant Banyal[1], *Mayank Saxena*[2], *and Deepak Kumar Sharma*[3]

[1]Department of Instrumentation and Control, Netaji Subhas University of Technology
(Formerly known as Netaji Subhas Institute of Technology), New Delhi, India

[2]Department of Electronics and Communications Engineering, Netaji Subhas University of Technology
(Formerly known as Netaji Subhas Institute of Technology), New Delhi, India

[3]Department of Information Technology, Netaji Subhas University of Technology
(Formerly known as Netaji Subhas Institute of Technology), New Delhi, India

CONTENTS

1.1 INTRODUCTION AND MOTIVATION

1.1.1 Internet of Things

The last two decades have been catalyzed by developments on a myriad of technological fronts and these developments have severely affected the way in which society functions. Technology has been increasingly integrated with our way of living and daily life, ranging from the moment we wake up at home and use smart home appliances to the usage of integrated technology in the workplace to health monitoring and analytics of our sleep. This development has asymmetrically changed the way industries perceive and use technology and with the incumbent developments they have been trying more and more to integrate them into their operations for efficiency. The reports suggest that the estimated count of connected IoT devices is set to rise to 50 billion by the end of this decade [1]. The ecosystem involves a myriad of elements such as: IoT devices, sensors, actuators, network elements (servers, routers etc.) and associated industrial machinery. In this pursuit of connecting conventional devices across networks and over the internet, the Internet of Things and Web of Things (WoT) have been pivotal in catalyzing and catering to this need. IoT as an emerging technology offers novel solutions and optimizing paradigms to both conventional and unconventional industrial operations. One such example of this innovative behavior is the case of innovative transportation in the field of Intelligent Transportation Systems (ITS) where IoT and associated technology have provided the ability for smart traffic management and traffic prediction through monitoring and predicting traffic location.

As discussed above, the Internet of Things or IoT encompasses a global network of nodes and devices that are addressable uniquely via standard communication protocols. The Internet of Things, which has witnessed a dramatic surge in the recent past, has had an immense impact on every aspect of human lives, ranging from wearable gear to sensors monitoring ecological changes in remote locations to regulating physical metrics in manufacturing processes. The set of devices or the "Things" which share a common resemblance in order to directly or indirectly connect to the Internet, operate within the confines of their functionality and exchange, analyze, process and deliver data in the common language; these sets of devices working in tandem are defined as "Internet of Things". Although large swathes in advancement in technology have unequivocally reduced human intervention and have significantly integrated devices with the real world, the big question of privacy and protection in IoT devices has been left largely unaddressed and now presents a potential threat to the cyber landscape. The lack of a standard IoT framework safeguarding privacy across all platforms has been attributed to varied communication protocols, a multitude of programming languages and differing levels of distributed computing in devices, networking and perceiving data in real-time systems [2].

The developers and research community have been meticulously working to develop tailor-made frameworks and structures for specific platforms. In its pursuit of this, the community has encountered several challenges pertaining to hardware which involve energy efficiency, ranging from the lightweight computation of devices and sensors to virtual threats including encryption attacks which occur on system vulnerabilities and tend to impede system integrity. Privacy is another concern that many nations across the globe have echoed. Policy measures such as the EU's General Data Protection Regulation (EU GDPR) have already been enforced with stringent rules for privacy yet there exist several challenges on the regulatory and technological front which this chapter touches upon in its first section. Industries such as healthcare, which incorporates one of the largest numbers of IoT devices, are especially under threat as revealed in the analysis by the Ponemon Institute and IBM. The most severe example of this is the case of Singapore, when an attack on SingHealth exposed the data of more than 1.5 million patients. The aforementioned cyber threats present us with a unique conundrum.

1.1.1.1 IoT Architecture

Every IoT system implemented globally is different; however, the data process flow and general architecture have some similarity. The first element is "Things"; this entails the nodes/devices that sense data from the environment via embedded sensors and actuators and are connected to the internet via appropriate gateways. The second layer includes the data acquisition systems and gateways that are responsible for gathering large amounts of raw and processed data (filtration, amplification and other associated electronic signal conditioning), and convert it into a digital form that is ready for further analysis. The third layer is where data visualization and intelligent control steps in, through which the processed data is transferred for long-term storage to data centers and cloud-based facilities which form the fourth layer.

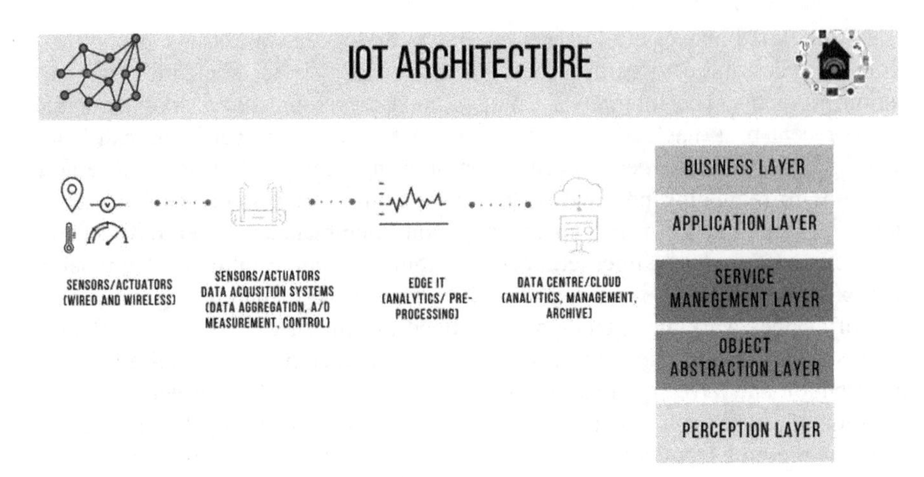

FIGURE 1.1 IoT general Architecture.

These four layers are illustrated in Figure 1.1 of this chapter. The figures entail a five-layer architecture that comprises of:

1. Business Layer
2. Application Layer
3. Service Management Layer
4. Object Abstraction Layer
5. Perception Layer

The business layer is responsible for the management of all activities, services and development of business models, graphs, and flowcharts based on the data it receives from the application layer. Further, this layer is responsible for supporting the decision-making aspect, based on big data analysis and determining the course of action. The application layer is responsible for service delivery and acts as an interface to the business layer. Furthermore, it is responsible for providing a control mechanism for accessing data and provides global management of the application based on objects' information processed in middleware. The service management layer is responsible for the pairing of services with their requester based on addresses and names, and for processing received data, making decisions and delivering the required services over network wire protocols. Furthermore, it is tasked to receive and process data from other layers. The object abstraction layer is responsible for the transfer of data produced by the objects to the service management layer. Also, it is responsible for transmitting data between devices and from devices to the receiver. Lastly, the object or perception layer is responsible for collecting sensor data in addition to digitizing and transferring data to the object abstraction layer. The details of the architecture and associated aspects are described in online literature [3–7].

TABLE 1.1
Vulnerability in IoT device

Assailability in IoT Device	Type of Vulnerability
Hardware layer	a) Lack of tamper resistance b) Weak embedded crypto algorithms c) Weak hardware implementations
Software layer	a) Firmware Layer b) Operating system c) Application layer
Communication protocols	a) Link & network layer protocol threats b) Application layer protocol threat c) Network design flaws
Key Management	a) Absence of support for public key exchange b) Easily extractable communication keys c) Employing of common or no key

Attacks that focus on IoT devices that have resource constraints have increased significantly in the past few years. The vulnerabilities in the security sector of the IoT technologies used are incessantly being identified; these technologies are used in both industrial and home environments such as sensors, industrial actuators, home appliances, medical devices, etc. The current state of affairs is exacerbated by defects in application, hardware chips that are faulty, and tamperable devices along with misconfigurations.

This section aims to use a risk-like approach to examine cyber attacks with respect to IoT-enabled devices, so as to highlight its existing threat landscape and isolate hidden and covert attack paths taken against critical infrastructure.

In IoT-enabled cyber attacks, the device is the amplifier or the enabler of an attack; the perpetrator identifies and takes advantage of inherent vulnerabilities related to one or multiple layers of the device so as to achieve his/her goal. We classify IoT vulnerabilities in two primary classes: "Embedded Vulnerabilities" and "Network Vulnerabilities".

1.1.1.2 Distinguishing IoT from Conventional Networks

Since its inception, IoT has experienced significant development in parallel to conventional networks. In comparison to the internet the network connection is established via physical links between web pages. Conventional networks are relatively more mature and well established on the technological front and can communicate via natural languages with efficacy. This is the reason which substantiates the prevalence of traditional networks and ease of their operation. In the IoT domain the standardization efforts are in their infancy and currently require skilled programming experts to implement an application.

1.1.2 Blockchain: An Overview

Blockchain is defined as a "public, permanent, appended-only distributed ledger" [8]. The issue of trust in information systems is extremely complex and quite prevalent.

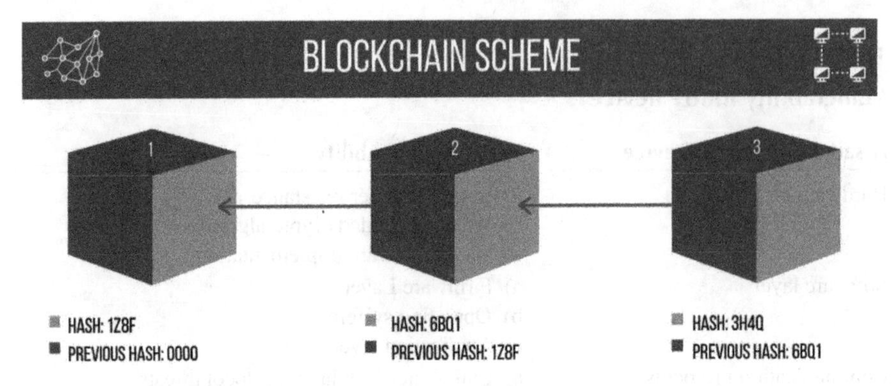

FIGURE 1.2 Blockchain Mechanism with use of reference Hash.

TABLE 1.2
Bitcoin node and functionality

	Wallet	Storage	Mining	Routing
Bitcoin Core	✓	✓	✓	✓
Full Node		✓		✓
Solo Miner		✓	✓	✓
Light Miner	✓			✓

This situation is exacerbated in the absence of audit and verification mechanisms, particularly in the case of systems handling sensitive information such as but not limited to financial and economic transactions. The problem of double spending is solved by enabling blockchain technology in a peer-to-peer network where there is an absence of a trust-based system. Blockchain enables verification of transactions by a group of unreliable actors. This aims to provide an immutable, distributed, secure, transparent and auditable ledger. The chain may be accessed openly allowing access to all transactions since the genesis transaction of the system. The protocol structures a chain of blocks that are linked to its previous block by a reference, thus forming a chain. Figure 1.2 describes the blockchain mechanism along with the use of a reference hash.

In order to support and operate the blockchain, network peers provide functionality which can include functions such as storage, wallet, and service and mining. Based on their functionality they can be part of different networks. Table 1.2 compares common types of nodes in bitcoin networks. Further, it does so establishing a consensus-based mechanism in which the nodes vote via their CPU power on the computation of a proof of work in the form of a hash for a given block which is based on the work that came previously.

1.1.3 Generations of Blockchain

Currently we are witnessing a critical shift toward distributed applications. This enables decentralized data sharing via secure transactions. This section reviews the emergence of blockchain in form generation starting from Blockchain 1.0 to Blockchain 4.0.

1.1.3.1 Blockchain 1.0: Bitcoin and Cryptocurrency

The first ever recognized generation of blockchain can be attributed to the rise of distributed ledger in form a virtual currency/coin, Bitcoin. The virtual coin enabled users to perform financial transactions over the internet. In addition, the currency is also referred to as "cryptocurrency" as it uses two keys to enable and authenticate the transaction:

Public Key: for verification of the legitimacy of the transaction
Private Key: for signing the transaction (enablement)

The Bitcoin ledger is composed of states of ownership of all existing bitcoin users informed of transactions between states, and output of any transaction state is essentially the transactional value if the transaction was successful. The copy of the above finite-length state transition system is maintained as a ledger record by the nodes of the network. The roles of third parties were eliminated in this decentralized and anonymous system as the proof of work is carried by hashing schemes based on Hashcash [9] and SHA-256 [10].

Figure 1.3 above illustrates the bitcoin transaction process, wherein the purchaser is referred by the entity to his signature which is a 16-digit encrypted code. The

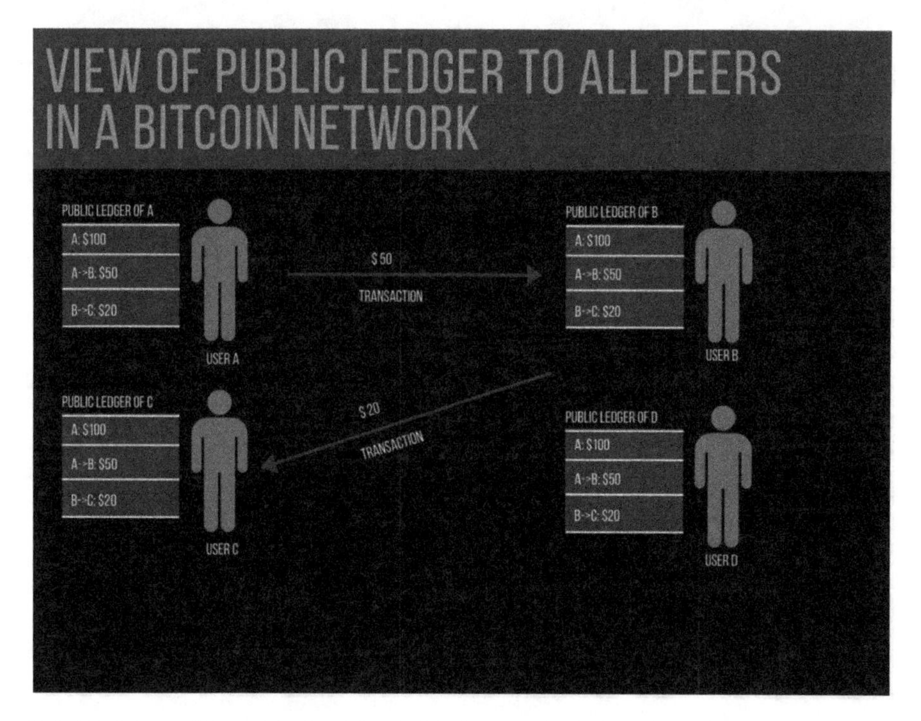

FIGURE 1.3 Public Ledger in a Bitcoin(BTC) network via State Transfer Function(STF).

signature is decoded by the purchaser at his receiving node, thus making the currency digital in nature over a decentralized and anonymous network.

1.1.3.2 Blockchain 2.0: Smart Contracts and Ethereum

The advent of Bitcoin (BTC) marked the rise of decentralization in computing, but the limited purview of BTC renders it unsuitable for general-purpose applications. This requirement of general application based systems was felt and in 2013 this was catered for to some extent with the launch of Ethereum. Ethereum is a blockchain coupled with an inbuilt Turing Complete programming language; this solved several scripting-based issues in BTC. This enabled users to create virtual ownership, the format for specific transactions and the state transfer function. This facilitated the growth of computer programs which existed and executed in a block chain—"Smart Contracts". These execute on their own in an autonomous manner through a set of predefined conditions. This resulted in reduction of the cost of verification and arbitration and enabled greater transparency in a transaction.

Figure 1.4 depicts the implementation of a smart contract on an Ethereum Blockchain. This includes a 20 B address and an STF. The contact code gets saved, authenticated and executed on a blockchain; each transaction comprises the following components:

1. Nonce
2. Ether Balance

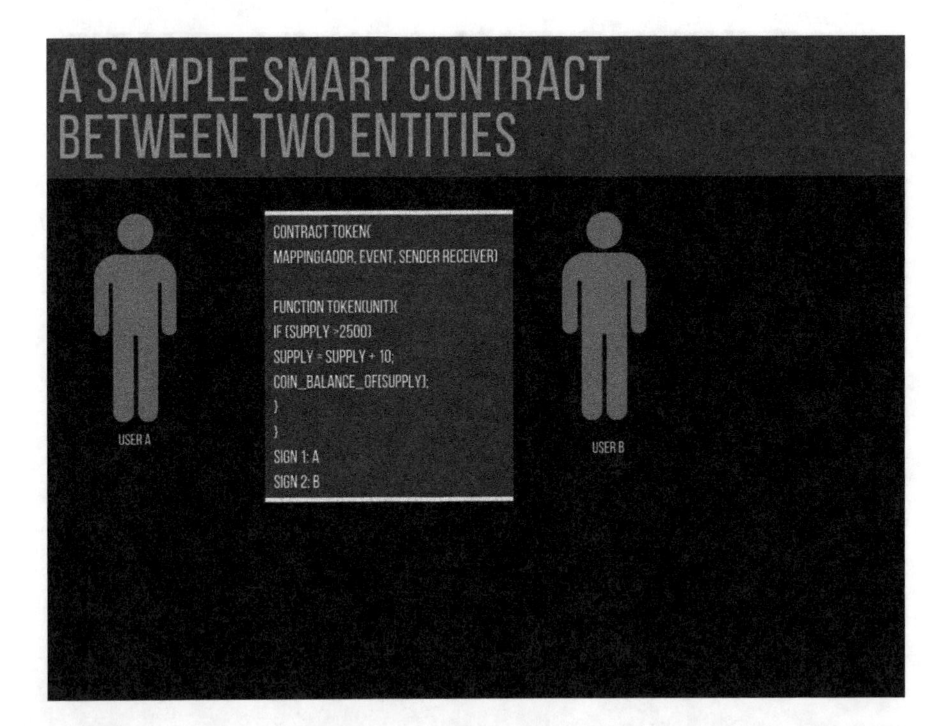

FIGURE 1.4 Illustration of a Smart Contract between two contracting entities.

3. Code Hash
4. Storage Root

1.1.3.3 Blockchain 3.0: Convergence toward Decentralized Applications

There is a lack of infrastructure evident as the existing technology is unable to sustain the volume of micro transactions with the prevalence of smart contracts. Consequently, there has been a shifting trend for blockchain toward decentralized networks and eventually a decentralized internet. This will integrate information storage, Smart Contract and communication networks. Thus there is a strong need for decentralized applications or D-App which have their backend enabled on Blockchain.

1.1.3.4 Blockchain 4.0: Seamless Integration with Industry 4.0

With the rise of decentralized applications there is a need for a common platform which will be a confluence of a myriad of applications and services that facilitate cross-platform communication. This enables entities to collaborate from distinct platforms to collate and work a single unit thus catering to the requirement of Industry 4.0. Industry 4.0 is used to label the trend in industry which emphasizes automation and confluence of cyber space with physical space in conjunction with IoT, Artificial Intelligence and cognitive computing.

1.2 IOT ARCHITECTURE AND SYSTEMIC CHALLENGES

1.2.1 Sensing Layer: Introduction and Challenges in End Nodes

Amidst the vast variety of IoT devices that surround humans, the most common are sensors, actuators, RFID readers, RFID tags, etc. These devices form a set of devices which are collectively termed as the sensing layer of IoT architecture. The critical contribution of this layer in IoT can be broadly summed up as sensing of ambient parameters and transmission of sensed data for processing in the next layers [8]. A few parameters that need to be considered in the sensing layer:

a) Cost, Resource and Energy consumption: The devices are equipped with minimal energy resources and memory in order to reduce cost.
b) Communication: The devices act as receiving ends of information and are designed to communicate with other devices on the network.
c) Networks: WSNs (Wireless Sensor Networks) and WMNs (Wireless Mesh Networks) connect a unique category of things in a complex; wireless and autonomous networks are employed for data acquisition, transmission and operation.

Figure 1.5 explains the fundamentals of service-oriented architecture in an IoT network and also its interaction with the other layers of IoT infrastructure.

Coupled with synchronized computing and communication capabilities, IoT is attributed to tapping into the potential offered by these individual sensors, turning them from classic to smart. In this regard, the security of the end nodes of the sensing layer of this network becomes of prime importance, particularly owing to

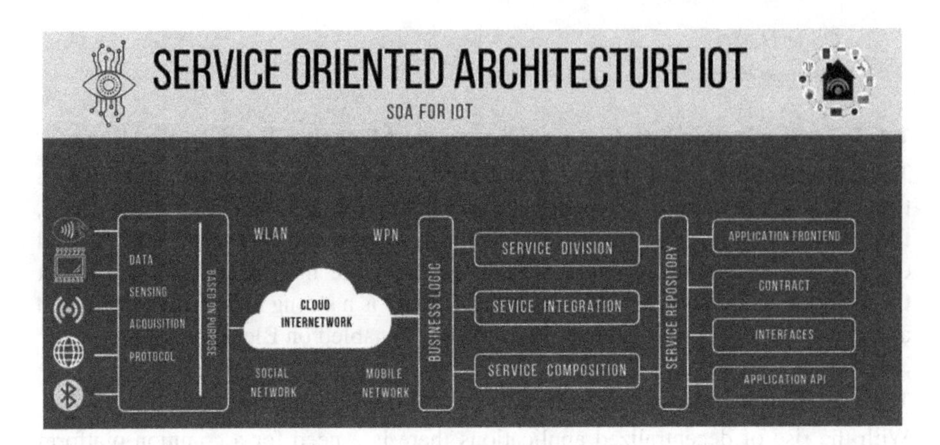

FIGURE 1.5 Service-Oriented Architecture IoT.

the uncertainty regarding data controlling. In this regard, the foremost prerequisite for the security mechanism in the Internet of Things is to have the rationale to take its own decisions which includes approving a command to accept, execute or terminate it. However, the confines of "Things" set up for minimal energy consumption and limited memory pose an extended range of security vulnerabilities at the sensing layer and end-node.

Upon classifying the various insecurities and threats that the sensing layer of IoT faces, it is essential to follow a few security preconditions; these include:

a) Security Prerequisites in IoT end nodes: confidentiality, integrity, privacy, access control, authentication, physical security protection and nonrepudiation.
b) Security Prerequisites in the IoT sensing layer: device authentication, authentication of information source, availability, integrity and confidentiality.

In order to achieve the above-mentioned requirements in the sensing layer of the IoT network, a few actions suggested include:

a) Creation of a trustworthy data sensing system and reinstating privacy and confidentiality of all devices in a network.
b) Identification of the source of users forensically as well as tracing them.
c) Designing the software or firmware at IoT to secure end nodes.
d) Administer security standards for all IoT devices.

1.2.2 Threat Based on Network Layer

For the optimum utilization of data procured in the sensing layer, it is equally important to transmit data among the IoT infrastructure. The network layer, therefore, provides the necessary medium to exchange information.

For smooth functioning and coordination among IoT devices, proper arrangement, organization and management of networks is important for which certain prerequisites include:

a) Effective network management such as wireless networks, fixed networks or mobile networks.
b) Energy efficiency within network layer.
c) QoS requirements.
d) Maintenance of privacy, confidentiality and security.
e) Mechanism for mining and searching.

Among the above-mentioned requirements, maintenance of privacy, confidentiality and security lies within the purview of the chapter and its importance is critical based on the complexity and mobility. Although existing security protocols and frameworks have provided security against threats and vulnerabilities until now, there exist a multitude of concerns that need to be addressed:

1) Broad security provisions: Provisions to ensure confidentiality, integrity and privacy for group authentication, protection of keys and availability of data.
2) Protection against privacy leakage: The location and complexity of certain devices in an IoT network often troubles the developers fearing the susceptibility of attacks upon sensitive data like user identity and credentials.
3) Secure communication: For an IoT system to exist, it must be fortified against attacks and reinforced with robustness, trustworthiness and confidentiality.
4) Fake network message: Creating fake signals corner and propagating miscommunication among devices from the entire network.
5) Overconnection: A highly connected network is also at risk due to two main reasons:

    a) High network congestion caused by signaling authentication on a bandwidth which may lead to a DoS attack.
    b) Intensive resource consumption caused by key operations and key security.

6) MITM attack: The attacks carried out independently by attackers over networks to forge a private connection while the attacker is controlling the entire conversation.

Although the innovations and technology available have been able to keep major threats at bay until recently, the growing influence of attackers has sent shockwaves across the globe. A series of steps in the following directions could help to provide greater security in the future:

a) Stringent authentication/ authorization process.
b) Secure transport encryption.

1.2.3 Service-Layer Based Threats

Upon sensing and transmission, the data procured requires operation while utilizing and integrating services of hardware and software platforms. The service layer hence aptly touted to be the middleware technology is designed according to the application requirements, application programming interface and service protocols to the standards of service providers, vendors and organizations. This layer is responsible for integration, analysis, security, management of UI and event-processing services [9] To render these services, the following steps are taken:

a) Service detection: Locating the optimum infrastructure necessary to conduct services efficiently;
b) Service combination and integration: To further broaden the scope of interaction among services and draw out more reliable ones is achieved by interaction by scheduling or recreating;
c) Authentication Management: Focus is laid upon verification of trusted devices through other services;
d) Service APIs: These help to improve interconnections between services.

To tackle the numerous challenges and threats, developers and corporations have contributed relentlessly to offer solutions to augment and improve services within the connected network. The ambitious SOCRADES integration architecture aims to ameliorate the interactions between application and service layers efficiently [11]. The "things" in the interconnection of devices are often limited to delivering services while exploiting these devices for discovery of networks, exchange of metadata and asynchronous publish and subscribe events [12]. In Peris-Lopez *et al.* (2006) [13], to increase the interoperability of loosely coupled devices and distributed applications, representational state transfer is set up. In Hernandez-Castro *et al.* (2013) [14], a service-provisioning process is introduced in the service layer that could strengthen ties and buttress cooperation between applications and services.

In light of the above-mentioned challenges and solutions offered to counter them, it is imperative to understand that certain security precautions, requirements and protocols, if undertaken, could shield against attacks in the service layer. A few of them are:

a) Dedicated authorization methods for service verification, authentication of groups, protection of privacy and integrity for the upkeep and storage of keys;
b) Protection against privacy leakage and location tracking;
c) Tracking services involving unauthorized use and unsubscribed services;
d) Prevention of potential threats like DoS attacks, node identification masquerade, replay attacks, service information manipulation and communication and services repudiation.

The solutions offered in this section broadly cover the solutions from major potential security threats.

In spite of the solutions offered, there remain a few open challenges that need to be addressed when creating an IoT application or services:

a) Securing data transference between layers;
b) Securing service management.

1.2.4 Application Interface Layer

This interface is the most visible and interactive layer of the IoT network and encompasses a myriad of utility-based implementations ranging from radio frequency identification based tracking to intelligent home management that are enabled via standardized protocols and other technologies [15]. Application maintenance requires certain security preconditions such as:

a) Safety-based isolation.
b) Secure methodologies for acquisition of software and updates.
c) Patches for augmenting security.
d) Verification means for the administrators.
e) Integrated platform for enhancement of security.

Different layers of IoT architecture require the following in order to sustain security in communication between the layers:

a) Maintaining the three tenets of security (privacy, confidentiality and integrity) for inter-layer communication.
b) Verification and approval of administrators cross layer.
c) Isolation of critical data.

The following regulations could prove to be helpful in designing security solutions:

a) The safety of these nodes should be attended carefully as most of the nodes in question are unsupervised.
b) Energy efficiency of nodes is of utmost importance while designing security solutions considering their large numbers.

1.2.5 Cross-Layer Challenges

Across all layers of IoT architecture through which the data is shared, certain standards are to be maintained to ensure that the network remains secure and fully interoperable. With the growing number of things in the network, it is the prerogative of the users to ascertain that their data is guaranteed protection against challenges among layers of the architecture.

The security needs across layers are virtually the amalgamation of the challenges faced across the IoT network:

a) Security protection in terms of design and execution time;

b) Ensuring high privacy standards to protect personal data through enhancement technologies;
c) Reinstating trust in IoT architecture

1.3 CHALLENGE TO IMPLEMENTATION OF BLOCKCHAIN IN IOT

1.3.1 Absence of IoT-Centric Consensus Protocol

A consensus protocol is a technique or a set of rules and regulations that make all the full nodes finalize on an arrangement over the sequence of transactions. There are various categories of consensus protocols being currently utilized in a variety of blockchain applications, for example, PoS, Proof of Work (PoW), Practical Byzantine Fault Tolerance (PBFT), etc. A few of the commonly used consensus protocols are deliberated upon in the following paragraphs of this chapter. Makdhoom [16] shows the contrariety of a few of the widely known and used blockchain consensus protocols in a thorough and exhaustive manner. The established consensus protocols such as Proof of Stake (PoS), Proof of Work (PoW), IOTA, Proof of Elapsed Time (PoET) are fabricated for permissionless blockchains, with an emphasis on financial transactions, whereas PoS and PoET have the potential to be used in permissioned blockchains.

The universal problem with these consensus protocols is the probabilistic nature of the consensus process and that it fails to terminate in a perdurable committed block. This further explains their vulnerability to blockchain forks [17]. One of the major contributors to deferred transaction verification is an absolute absence of consensus finality that is not at all sustainable for the various real / near-real-time IoT systems that demand instantaneous transaction verification and completion. Furthermore, examining the various consensus protocols, we find that a certain type of hardware is required by PoET in which the region allocating wait time necessitates being a trusted entity.

A number of ambiguities are yet to be resolved when it comes to IOTA since it is still in the open beta testing phase particularly regarding its performance efficacy and security. A number of questions pertaining to whether IOTA would be an effective micro-payment method or its compatibility with contracts as in HFB (Hyperledger-Fabric blockchains) and Ethereum and its capability of providing data confidentiality are still unanswered.

Meanwhile, a number of other consensus protocols like the Delegated Byzantine Fault Tolerance (DBFT), Practical Byzantine Fault Tolerance (PBFT), HoneyBadger *et al.* are based on the Byzantine Fault Tolerance (BFT) protocol which is a group of state machine reduplication protocols. By replicating the services on a multitude of nodes it provides security against arbitrary faults. Despite BFT being the preferred protocol for permissioned-type blockchains, it has several shortcomings. BFT-based protocols have a high vulnerability toward DoS attacks owing to their languid timing assumptions with the exception of HoneyBadger BFT [18]. Not only does the feeble synchrony negatively alters the system's productivity, but the liveness characteristic of languid synchronous protocols is also unsuccessful as the feeble timing assumptions get contravened when vitriolic networks launch DoS attacks.

Scalability pertaining to the number of validator nodes is another challenge that comes with BFT-based protocols as the testing mechanism is not usually implemented post 20 nodes [19]. This shortcoming can be overcome by executing Algorand [20] that implements a system of random selection of a small-sized committee for every step of the consensus protocol to address scalability. This random selection is done through the mechanism of Verifiable Random Functions (VRF). Committee size is contingent on two constraints in Algorand:

i. $1/2g + b \leq Tstep.\tau step$
ii. $g > Tstep.\tau step$

Where

g = number of honest committee members;
b = number of malicious committee members;
T = number of votes required for consensus;
τ = expected committee size.

Furthermore, BFT-based consensus protocols are proficient in obscuring non-deterministic faults that eventuate at $f = (n-1)/3$ replicas, where

f = number of faulty nodes
n = number of total nodes

Proof of Work (PoW) and BFT-based consensus protocols differ largely on the basis of availability which is an important necessity in IoT systems. On one hand where PoW does not determine whether a certain unresolved transaction will be present in the next bloc or not, PBFT on the other turns out to be particularly exorbitant when it comes to message complexity. Hence the sustainability of a huge number of IoT systems should be the focus of any future solution that is based on blockchain and should also be in accordance with the wireless communication rules of the particular country.

Transaction validation rules, fault tolerance and transmission complexity are some of the aspects that need to be enhanced if consensus-based protocols are to be applied in IoT-based systems.

1.3.2 Transaction Validation Rules

Figure 1.6 represents the transaction verification and completion procedure in bitcoin. The working protocol for this functions on a specific set of regulations encompassing the authentication of transaction format and correct signatures and then followed by a mechanism that checks whether the specified transaction has been previously spent or not [21]. In comparison, Ethereum has a different process for transaction validation. As Figure 1.7 depicts, Ethereum verifies signatures, gas, nonce, account balance of sender's account and the format.

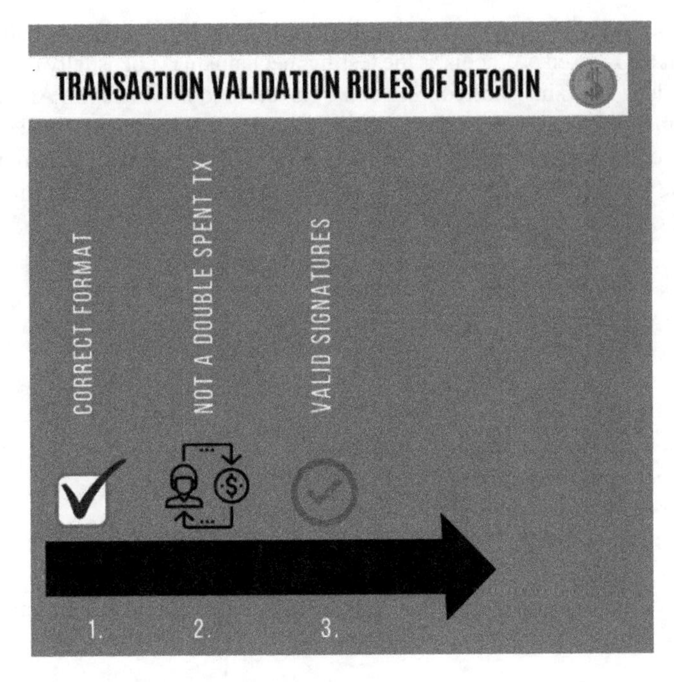

FIGURE 1.6 Transaction validation rules in Bitcoin.

Due to the vulnerability of IoT systems toward cyber attacks, it raises an important conundrum as to whether or not the present regulations for transaction verification of blockchain are compatible with the IoT system. This problem is further augmented due to the fact that IoT systems, for the most part, consist of heterogeneous devices and applications that send data in a distinct range of values. These devices being highly vulnerable can be infected even through generic malware attacks that would consequently also be used by a botnet for even more attacks. Hence, the transaction verification regulations of blockchain may not be appropriate for IoT systems.

1.3.3 Scalability Challenges

When it comes to discussing the challenges with respect to the scalability of integrating blockchain in IoT systems, two major issues arise—storage capacity and the inherent latency of blockchain. The scalability affects both the size and the consensus procedure. For instance, if the number of users increases it will directly result in the rise of transactions as well. Consequently, the consensus protocol affects the delay in transaction validation. These challenges are discussed at length in the following paragraphs of this chapter.

1.3.3.1 Storage Capacity

Typically, blockchain does not have the capacity to contain a huge amount of information, whereas a smart city IoT system containing hundreds and hundreds of end nodes has the ability to produce a large amount of data in a very small duration

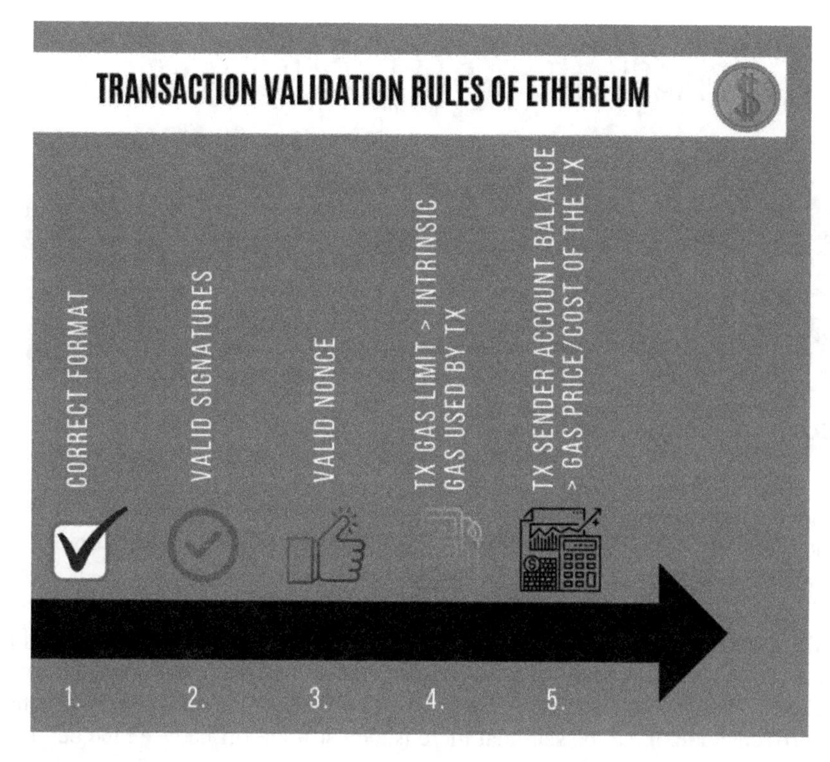

FIGURE 1.7 Transaction Validation Rules Ethereum.

of time which needs to be processed to extricate data for a multitude of uses. The integrability of IoT devices is further limited due to the necessity of storing the entire blockchain through the full and miner nodes. Furthermore, the impediments on resource-limited devices to perform as validator nodes is augmented with the perpetual expansion in blockchain's size which subsequently increases the storage necessities. The synchronization time rises as well, as each new user/device is added to the network. Hence it becomes extremely challenging to produce such a mechanism that encompasses both the constraint resources of IoT and the advantages of blockchain.

1.3.3.2 Inherent Latency Blockchain

The requirement for an enhancement in the transaction verification duration of real-time IoT systems like Industrial Control System (ICS), smart vehicles, Wireless Sensor Network (WSN), intelligent transportation systems, *et al.* without making concessions on its safety and efficiency has risen in demand.

For instance in the Proof of Work (PoW) based blockchain, the transaction verification duration is reduced if the block production duration lessens. However, a transaction needs to wait for more verifications to realize the identical strength of security because of the reduced complications in mining the block. Furthermore, a rise in the dissipation of computational resources is incurred due to the increase in the stability

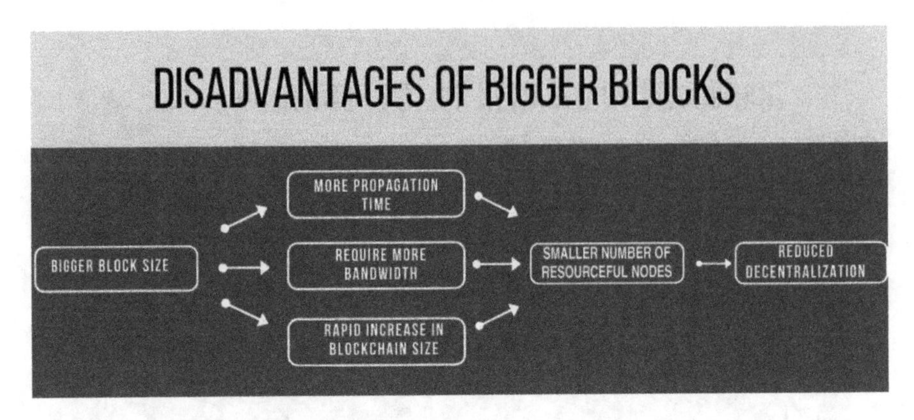

FIGURE 1.8 Disadvantages of Bigger Block.

of blocks. It is believed that the throughput of a bitcoin blockchain can be augmented if the block size compounds from 1 MB to 2 MB. However, it is the larger block that utilizes a larger duration of time to transmit in the network.

Figure 1.8 illustrates the disadvantages of implementing bigger blocks. Furthermore, the number of full nodes is influenced by the expansion in the block size as an increased amount of resources are hence necessitated to store the entire blockchain which consequently factors into a swift development of the blockchain size. To conclude, it can be said that there is a clear trade-off between the degree of reliability and execution efficacy. A distributed network will have performance issues and a centralized network will harbor trust issues.

1.3.4 IoT Device Integration Challenges

Even though only Ethereum and Hyperledger-Fabric are a few of the only blockchain techs that incorporate smart contracts and distributed applications, they have a major shortcoming vis-à-vis execution of smart contracts in an Ethereum Virtual Machine (eVM) does not transmit in a direct fashion with the rest of the world. Hence, we make use of the web3.js library as an interfacing mechanism. Consequently, a safe decentralized database is one of the useful utilizations of blockchain.

As the present situation is with regard to the threats in IoT devices as they can be undermined without much difficulty, the integrity of IoT systems will always remain a matter of concern. Furthermore, the data can be further compromised due to an infectious code implemented remotely or simply due to a software/hardware malfunction or even human interference. These errors are not determined by the system unless they are specifically tested for such kinds of failures, modifications or misconfigurations. At present, the sole accessible solution is Orcalize [22] which takes information from third-party sources like IPFS and WolframAlpha. The provision of a Proof-Of-Authentication is mandated to establish the authenticity of the data provided.

Unfortunately, Orcalize is not compatible with IoT devices and hence their integration becomes even more difficult. Therefore, the requirement of an additional client-interfacing software between blockchain and the IoT device will introduce more computational and memory expenditure. Hence, a lot of attention needs to be paid to ensure that an IoT device can have a vast variety of interchanges with blockchain.

1.3.5 Protection of Devices against Malware and Content Execution Attacks

The protection of devices against malware and content execution attacks has a dual aspect to it—ransomware and malware. Ransomware, however, has minimal effect on a distributed ledger as the network contains the error-free copy of the ledger even though some of the nodes may get affected. On the other hand, malware has the capacity to infiltrate the network with counterfeit/unauthentic data through a compromised node. Since the sensors have situation-based data which is challenging to authenticate with older transactions as opposed to in bitcoin, it becomes extremely difficult for the nodes to verify any transaction/data. Henceforth, for the detection of malicious and counterfeit nodes, the existence of malware detection software is a necessity in such blockchain-based IoT systems.

1.3.6 Secure and Synchronized Software Updates

IoT devices and their applications continue to be in operation for an extended period of time due to their analytical functionalities with the absence of any software or firmware upgrades and hence are very prone to cyber attacks. Even though because of the localized and distributed framework of blockchain, at present, synchronized firmware and software updates cannot be assured, the need for such a mechanism is increasing day by day.

1.4 APPLICATION OF BLOCKCHAIN IN IOT SECTOR

1.4.1 Autonomous Decentralized Peer-to-Peer Telemetry

In 2015, the International Business Machines Corporation released the Proof of Concept (PoC) for a Decentralized Peer-to-Peer Telemetry System (DePT) that is autonomous in its functionality in order to utilize blockchain's ability to perform smart contracts on the verification of transactions [23].

The objective of DePT is to execute and administer a localized, distributed, safe, independent, expandable and powerful architecture for IoT that does not contain singular vulnerable failure functions. The suggested architecture makes use of the TeleHash protocol for peer-to-peer data transfer and BitTorrent as the mechanism for distributed sharing. This suggested system endeavors to solve several issues pertaining to the traditional IoT systems concerning privacy, failure points, safety regarding centralized entity and problems introduced due to human interference. An autonomous decentralized telemetry system also aims to proffer user and data protection, identification management and the peer-controlled access to data.

A major void that exists with regards to its implementation is that this system is a Proof of Concept (PoC) and hence it needs testing and development to determine its dependability pertaining to safety and functioning efficacy.

1.4.2 Blockchain-Enabled Security for Smart Cities

A multitude of challenges regarding the problems faced during data sharing from heterogeneous applications occurs due to the non-existence of a universal protocol for smart devices in a traditional environment. This further hampers their integration and the provision for cross-functionality characteristics. Examining the synopsis of a security system based on blockchain for reliable transmission among smart city units presented in Muthukkumarasamy and Biswas [24] states that smart cities aiming to provide a shared environment for secure transmission require integration of its devices with blockchain. Furthermore, an incorruptible log of transfers will be available for auditing purposes due to the use of blockchain in such a system.

At present, there does not seem to be a computational inclusive quantitative and qualitative analysis of blockchain-enabled smart city entities; however, it is quite ambiguous which platform, transaction technique or consensus protocol will be the most suitable for efficient implementation.

1.4.3 Blockchain-Enabled Smart Home Architecture

A safe, personal and lightweight framework for Smart Home Applications based on blockchain has been advocated by A. Dorri, S. Kanhere and R. Jurdak in [25] and [26]. What we need to understand here is that a Smart Home blockchain differs from a traditional Bitcoin blockchain in a multitude of ways. The owner operates the smart home blockchain alone as opposed to in bitcoin and hence, the owner has complete control over the transactions inside his Smart Home. Furthermore, it also proffers limited access to IoT data that guarantees information integrity, availability and confidentiality and also safeguards against DDoS threats. A blockchain-enabled Smart Home architecture intends to solve other challenges like the strenuous computations, latent transaction verification and power consumption through imitating Proof Of Work (PoW) in the process of block mining. The suggested framework makes use of cloud storage space to reduce the strain of memory necessities on smart home appliances.

There are a number of problems when it comes to the implementation of a Blockchain-enabled Smart Home. The distinguishing feature of blockchain is its distributed network but in a smart home environment, the Home-Miner or the owner solely has control over the entire network. This introduces a single juncture failure at both the Home-Miner level and the cloud storage space level. Secondly, the absence of a verification mechanism on a consensus basis as the home-miner has complete control. Furthermore, if the home-miner itself is malicious or corrupted, then the integrity of the transactions cannot be determined. Also, in this case, the home-miner determines if a new block will be added whereas, in the conventional blockchain, it is a consensus decision.

1.4.4 Blockchain-Based Self-Managed VaNeTs

In order to address the challenges posed by a centralized traditional Vehicle Ad-Hoc Network (VaNeT), such as the single juncture failure and vulnerability toward attackers and reduced user privacy, Leiding [18] has advocated for a distributed Self-Managing VaNeT based on Ethereum Blockchain that has a challenge-response verification mechanism. The entire framework is run by appliances based on Ethereum that administer the regulations for proffering a multitude of services. The identification mechanism for every node is its Ethereum address. Each node makes a payment in ethers if it intends to use any Ethereum-based service which in turn becomes a funding mechanism for the network platform. This funding provides the necessary incentive for various merchants to continue manufacturing such applications and services based on Ethereum. Such an Ethereum account has the potential to make self-operating transactions; for insurance, registration and fines in case of violation of traffic rules.

A number of dilemmas are still unanswered pertaining to this suggested framework such as—What kind of data will be visible on the blockchain, Who will be mining the said block, What shall be the mechanism for V2V transmission and the inherent latency in it. An important point to be noted is that latency is an intrinsic characteristic of the blockchain technology but in traffic and road situations, real-time information is of paramount significance for nodes connected to a VaNeT.

1.4.5 Security and Privacy of Data

Nathan and Zyskind [27] propose a data management prototype for a distributed network that provides protection and safeguard measures on problems related to data proprietorship, transparency and auditability. Ethos is a bitcoin-based system for transmission of personal information manufactured by Viral Communications, MIT Media Lab [28]. Ethos's compatibility for its use in IoT systems is something that presently still needs evaluation. A distributed computational protocol named Enigma has also been developed with the purpose of preserving privacy [29]. Enigma further allows very limited access to the entire data by its nodes through the deployment of a multi-entity computation that has a secret-sharing validation protocol. This type of system has the added advantage of decentralized storage and reduced memory necessity for embedded devices.

For its efficacious execution in IoT systems, Enigma still necessitates analyses for the overhead transmission and computation. It is important that any solution for decentralized computation and safe data transmission are in accordance with the respective country's law for wireless communication since most of the IoT devices transmit through wireless media only. Even though such decentralized computational schemes seem quite efficacious, their productivity with respect to bandwidth efficiency still requires evaluation. In conclusion, any future data transmission and sharing systems based on blockchain should keep in mind these shortcomings.

1.5 CONCLUSION AND FUTURE SCOPE OF WORK

Technologies such as blockchain have disrupted the entire fintech market and even though it has created a lot of controversies, the technology is going to get more and more integrated into our lives. The advantages of integrating blockchain with IoT should hence be assessed very carefully and with the utmost caution, because without determining its risks and applying it in situations where the costs do not overpower the improvements is an easy trap to fall into. This chapter summarizes the challenges that come with blockchain and IoT and hence collective work must be done in order to address these problems. We have also been able to identify where this technology has the potential to enhance IoT applications.

Furthermore, we have provided a feasibility check on using the blockchain technology with IoT appliances where existing frameworks have been analyzed, conditions for future solutions identified and the current scenario of the blockchain-IoT paradigm exhaustively addressed. Moreover, special caution has been put on the need for future solutions to be contingent with the respective country's laws. This integration of technologies should hence become part of the government's framework which would subsequently speed up interactions with citizens as well.

The dual aspect of data integrity and framework supervision is of utmost importance. The privacy of users and security concerns affect how the citizens will perceive this integration. The challenges of storage space, scalability and consensus protocols will play an integral role in how this process moves forward. This integration of IoT devices with blockchain technology is bound to exponentially increase the applications and use of blockchain and establish its prominence at a similar level as in the current fiduciary money market.

REFERENCES

1. Al-Fuqaha, Ala, Mohsen Guizani, Mehdi Mohammadi, Mohammed Aledhari, and Moussa Ayyash. 2015. "Internet Of Things: A Survey On Enabling Technologies, Protocols, and Applications". *IEEE Communications Surveys & Tutorials* 17 (4): 2347–2376. doi:10.1109/comst.2015.2444095.
2. R. van Kranenburg. *The Internet of Things: A Critique of Ambient Technol- ogy and the All-Seeing Network of RFID*. Amsterdam, The Netherlands: Institute of Network Cultures, 2007.
3. A. Al-Fuqaha, M. Guizani, M. Mohammadi, M. Aledhari, M. Ayyash, Internet of Things: A Survey on Enabling Technologies, Protocols, and Applications, *IEEE Communications Surveys & Tutorials*, 17 (4) (2015) 2347–2376.
4. S. A. Kumar, T. Vealey, H. Srivastava, Security in Internet of Things: Challenges, Solutions and Future Directions, in: Proceedings of the IEEE 49th Hawaii International Conference on System Sciences (HICSS), 2016, pp. 5772–5781.
5. M. Khari, M. Kumar, S. Vij, P. Pandey, Vaishali, Internet of Things: Proposed Security Aspects for Digitizing the World, in: Proceedings of the 3rd International Conference on Computing for Sustainable Global Development (INDIACom), 2016, pp. 2165–2170.
6. R. Khan, S. U. Khan, R. Zaheer, S. Khan, Future Internet: The Internet Of Things Architecture, Possible Applications and Key Challenges, in: Proceedings of the IEEE 10th International Conference on Frontiers of Information Technology (FIT), 2012, pp. 257–260.
7. T. Qiu, N. Chen, K. Li, M. Atiquzzaman, W. Zhao, How Can Heterogeneous Internet of Things Build Our Future: A Survey, IEEE Communications Surveys & Tutorials.

8. Editors, MIT. 2019. "Explainer: What Is A Blockchain?". *MIT Technology Review*. www.technologyreview.com/s/610833/explainer-what-is-a-blockchain/.
9. A. Back, "Hashcash – A Denial of Service Counter- Measure," 2002, available at: www.hashcash.org/papers/hashcash.pdf
10. D. Eastlake, 3rd and T. Hansen, "US Secure Hash Algorithms (SHA and SHA-based HMAC and HKDF)," RFC 6234 (Informational), May 2011, available at: www.ietf.org/rfc/rfc6234.txt.
11. Choi, J., Li, S., Wang, X., Ha, J., 2012. A General Distributed Consensus Algorithm For Wireless Sensor Networks. Paper presented at the Wireless Advanced (WiAd), 2012
12. Fielding, R. and Taylor, R. (2002). Principled Design of the Modern Web Architecture. *ACM Transactions on Internet Technology*, 2(2), pp.115–150.
13. Kranenburg, R.V., Anzelmo, E., Bassi, A., Caprio, D., Dodson, S., Ratto, M., 2011. The Internet of Things. Paper presented at the 1st Berlin Symposium on Internet and Society (Versión electrónica). Consultado el.
14. Peris-Lopez, P., Hernandez-Castro, J.C., Estevez-Tapiador, J.M., Ribagorda, A., 2006. M2ap: A Minimalist Mutual-Authentication Protocol for Low-Cost RFID Tags. Ubiquitous Intelligence and Computing. Springer, Heidelberg, pp. 912923.
15. Hernandez-Castro, J.C., Tapiador, J.M.E., Peris-Lopez, P., Li, T., Quisquater, J.-J., 2013. Cryptanalysis of the SASI Ultra-Light Weight RFID Authentication Protocol. arxiv.
16. I. Makhdoom, M. Abolhasan, H. Abbas and W. Ni, Blockchain's adoption in IoT: The challenges, and a way forward in: Journal of Network and Computer Applications, 2019, vol. 125, pp. 251-279.
17. EconoTimes, Blockchain project antshares explains reasons for choosing dbft over pow and pos (2017)
18. A. Miller, Y. Xia, K. Croman, E. Shi, D. Song, The Honey Badger of BFT Protocols, in: Proceedings of the ACM SIGSAC Conference on Computer and Communications Security, ACM, 2016, pp. 31–42.
19. M. Vukolic, The Quest for Scalable Blockchain Fabric: Proof-of-work vs. BFT Replication, in: Proceedings of the International Workshop on Open Problems in Network Security, Springer, 2015, pp. 112–125.
20. Y. Gilad, R. Hemo, S. Micali, G. Vlachos, N. Zeldovich, Algorand: Scaling Byzantine Agreements for Cryptocurrencies, in: Proceedings of the 26th Symposium on Operating Systems Principles, ACM, 2017, pp. 51–68.
21. V. Buterin, et al., A next-generation smart contract and decentralized application platform, white paper.
22. Oraclize Is Now Provable Things". 2019. *Oraclize.It*. www.oraclize.it.
23. 2019.*Cryptonomics.Show*.https://cryptonomics.show/wp-content/uploads/2018/08/IBM-ADEPT-Practictioner-Perspective-Pre-Publication-Draft-7-Jan-2015.pdf.
24. K. Biswas, V. Muthukkumarasamy, Securing smart cities using blockchain technology, in: Proceedings of the IEEE 14th International Conference on Smart City High Performance Computing and Communications, 2016, pp. 1392–1393.
25. A. Dorri, S. S. Kanhere, R. Jurdak, Blockchain in internet of things: Challenges and solutions, arXiv preprint arXiv:1608.05187.
26. A. Dorri, S. Kanhere, R. Jurdak, P. Gauravaram, Blockchain for IOT Security and Privacy: The Case Study of a Smart Home, in: Proceedings of the IEEE 2nd Workshop on security, privacy, and trust in the Internet of things (PERCOM), Hawaii, USA, 2017.

27. G. Zyskind, O. Nathan, et al., Decentralizing Privacy: Using Blockchain to Protect Personal Data, in: Proceedings of the IEEE Security and Privacy Workshops (SPW), 2015, pp. 180–184.
28. MIT-Media-Lab, Ethos (2014. Last accessed 26 July 2018). URL http://viral.media.mit.edu/projects/ethos/
29. G. Zyskind, O. Nathan, A. Pentland, Enigma: Decentralized Computation Platform with Guaranteed Privacy, CoRR abs/1506.03471.

2 Application and Challenges of IoT in Healthcare

[1]Subhashree Sahoo, [2]Debabrata Dansana, and [2]Brojo Kishore Mishra
[1]Pondicherry University, Kalapet, Pondicherry, India
[2]GIET University, Gunpur, India

CONTENTS

2.1 INTRODUCTION

Healthcare changes dramatically as a result of technological developments, from anesthetics and antibiotics to magnetic resonance imaging scanners and radiation therapy. Future technological innovation is going to keep remodeling healthcare, however, whereas technologies (new drugs and treatments, new devices, new social media support for healthcare, etc) can drive innovation, human factors can remain one of the persistent limitations of breakthrough. No predictions will satisfy everybody. There are no two ways regarding healthcare; technological developments in healthcare have saved innumerable patients and are continuously raising our quality of life. Not solely that, however; technology within the medical field has had a huge impact on nearly all processes and practices of healthcare professionals.

Advancements in medical technology have allowed physicians to better diagnose and treat their patients since the start of the professional practice of medicine. Because of the continual development of technology within the medical field, many lives are saved and also overall quality of life continues to increase over time. Although medical culture is comparable, there are dramatic technological changes, and truly, these changes would be exhausting to explain. Does anybody even know how an infusion pump works? They used to be clockwork and currently nearly everything contains a computer and includes a color screen and plenty of buttons. Implanted defibrillators that use telephone networks and internet sites to keep cardiologists up thus far with their patients are simply magic. New pharmaceuticals that change moods, change pressure, or kill bacteria: all are trendy magic. Some of what appears to us these days like science fiction becomes routine in the future, even perhaps in our lifetimes. However, much of today's human story regarding relationships, hopes, error, grief, and denial is going to stay entirely recognizable within the future. We will still have authority gradients, we will still have dispute over human error, and patients can still be made helpless so they are easier to treat. The explanation is that technology is driven by the market.

If someone has an idea that they might convert into a physical realization that they can sell, they will additionally patent it or license it, and thereby build a return on their investment. This, in turn, can encourage them to seek out ways of making it smaller and cheaper, and selling it on a bigger scale. Thus it is technology-driven. In contrast, human culture does not build a profit for anybody.

The main reason behind healthcare is patients, and they should be at the center of it. In this article, we will discuss some of the technological trends and their challenges in healthcare. In this era of science, healthcare is nothing but a market for technology and in this case hospitals which act as consumers, and these consumers are ready to pay enough money for technological equipment which includes minimization of cost, division of labor and so on.

2.2 MEDICINE AND TECHNOLOGY

In today's world, technology plays a vital role in each business more than in our personal lives. Out of all of the industries in which technology plays a vital role, healthcare is certainly one of the foremost. This is answerable for saving innumerable lives all around the world.

Medical technology could be a broad field where innovation plays a vital role in sustaining health. Areas such as biotechnology, pharmaceuticals, information technology, the development of medical devices and instrumentation, and others, have all created vital contributions to increasing the health of individuals all around the world. From "small" innovations like adhesive bandages and articulatio-talocruralis braces to larger, additional advanced technologies like magnetic resonance imaging machines, artificial organs, and robotic prosthetic limbs, technology has beyond question created an unimaginable impact on medicine. In the healthcare trade, the dependence on medical technology cannot be exaggerated, and as a result of the development of those brilliant innovations, healthcare practitioners will still find ways to enhance their practice—from better diagnosing, surgical procedures, and improved patient care.

2.2.1 Information Technology and Medicine

Information technology has created vital contributions to our world, particularly within the medical trade. With the increased use of electronic medical records (EMR), telehealth services, and mobile technologies such as tablets and smart phones, physicians and patients are each seeing the advantages that these new medical technologies are bringing.

Medical technology has evolved from introducing doctors to new instrumentation to use within private practices and hospitals to connecting patients and doctors thousands of miles away through telecommunications. It is not uncommon in today's world for patients to conduct video conferences with physicians to save lots of time and cash that would otherwise be spent on traveling to a different geographic location, or send health information instantly to any specialist or doctor in the world.

With more and more hospitals and practices using medical technology such as mobile devices at work, physicians currently have access to any type of information they have—from drug information, analysis and studies, patient history or records, and so on—in mere seconds. And, with the power to effortlessly carry these mobile devices around with them throughout the day, they are never far away from the information they have. Applications that aid in distinguishing potential health threats and examining digital information such as x-rays and CT scans additionally contribute to the advantages that information technology brings to medicine.

2.2.2 Medical Equipment Technology

Improving the standard of life is one of the greatest advantages of integration new innovations into medicine. Medical technologies such as minimally invasive surgeries, better monitoring systems, and easier scanning instrumentation are permitting patients to pay less time in recovery and enjoy a healthy life for longer.

The integration of medical instrumentation technology and telehealth has also created robotic surgeries, where in some cases physicians do not even have to be compelled to be within the operating theater with a patient as surgery is performed. Instead, surgeons will operate out of their "home base", and patients will have the procedure carried out in a hospital or clinic in their own hometown, eliminating the hassles and stress of health-related travel. With different robotic surgeries, the doctor remains within the room, operating the robotic devices; however, the technology allows for a minimally invasive procedure that leaves patients with less scarring and considerably less recovery time.

2.2.3 Technology and Medical Research

Medical scientists and physicians are perpetually conducting research and testing new procedures to help prevent, diagnose, and cure diseases as well as developing new medicines that may reduce symptoms or treat ailments.

Through the utilization of technology in medical analysis, scientists are able to examine diseases on a cellular level and manufacture antibodies against them. These vaccines against dangerous diseases like protozoa infection, polio, MMR and others prevent the unfolding of disease and save thousands of lives all around the globe. In fact, the World Health Organization estimates that vaccines save about three million lives each year, and prevent lots of others from acquiring deadly viruses and diseases.

2.2.4 3D Printing

Today, it is possible to reproduce bones and a few internal organs using 3D printing technology. These artificial organs and bones will then be introduced into the body of the patient to replace diseased or problematic areas. Surgeons are also using

3D printing technology to gain a far better understanding of what is happening within their patients' bodies. With a 3D model, it is considerably easier for a surgeon to carry out a more in-depth check of the problem and simulate a variety of solutions or possible operations that may be undertaken before performing the actual surgery on the patient.

Similarly, 3D printing has revolutionized medical specialty. With a 3D printer, obtaining a custom-made prosthetic hand or leg is considerably cheaper. It is currently possible to custom print prosthetic hands, for example, for a baby that needs totally different models as it grows, rather than having to travel and get a replacement prosthetic hand fitted every year. Plus, with the huge developments that are being created within the 3D printing industry, prices related to this technology are reducing every day.

One of the various forms of 3D printing that is utilized in the medical device field is bioprinting. Instead of printing using plastic or metal, bio printers use a computer-guided pipet to layer living cells, described as bio-ink, on top of one another to make artificial living tissue in a laboratory.

These tissue constructs or organoids can be used for medical analysis as they mimic organs on a miniature scale. They are also being trialed as cheaper alternatives to human organ transplants.

Another application of 3D printing within the medical field is making patient-specific organ replicas that surgeons can use to practice on before performing sophisticated operations. This method has been tried in order to speed up procedures and minimize trauma for patients.

This type of procedure has been performed with success in surgeries starting from a full-face transplant to spinal procedures and is starting to become routine practice.

Sterile surgical instruments, such as forceps, hemostats, surgical knife handles, and clamps, are often made using 3D printers. Not only will 3D printing manufacture sterile tools; some are provided by the traditional Japanese practice of origami that means they are precise and might be created extremely small in size. These instruments are often used to operate on small areas while not inflicting inessential extra harm to the patient.

One of the main aspects of using 3D printing instead of traditional production methods to provide surgical instruments is that production prices are considerably lower.

3D printing within the medical field may be used to manufacture prosthetic limbs that are custom-made to suit and match the user. It is common for amputees to wait weeks or months to receive prosthetics through the normal route; however, 3D printing considerably accelerates the method, making less expensive products that provide patients continuing functionality just as traditionally factory-made prosthetics.

2.2.5 Digitization of Health Records

Electronic Health Records (EHRs) replacing outdated paper records has been a vast game changer for everybody within the medical world. Medical assistants, medical coding professionals, and registered nurses are just some of the roles that are covered by this industry-wide implementation.

Nurses and technicians are answerable for inputting patient information into a central, digitized system. Medical billers and coders update patient records with diagnostic codes such as test results and submit medical claims to insurance corporations. Not only will patients access their records at the click of a button; it is equally ensured that mistakes are caught a lot more quickly without needing to pore over unreadable physicians' handwriting.

In medicine, the primary information technology wave to hit the art and science of healing was the digitization of medical files, currently referred to as electronic health records (EHRs). The data contained in EHRs together with alternative sources has the potential to remodel medical practice by leveraging information, technologies, and healthcare delivery to improve the overall potency and quality of care at a reasonable price. The widespread adoption of EHRs has generated massive sets of information. The skillful merging of datasets collected from patients and physicians might be a viable avenue to strengthen healthcare delivery. These huge datasets are currently taken as a byproduct of medical practice instead of as useful assets that would play important roles in patient care. Currently, for example, most EHRs collect quantitative, qualitative, and transactional information; all of that could be collated, analyzed, and applied using sophisticated procedures and techniques that are currently available to make use of text-based documents containing disparate and unstructured data. The purposeful use of information is not a mystery to medical practice. From their humble beginnings, evidence-based undertakings are grounded within the principle that questions answered through the methodology were superior to anecdotes, expert opinion, panels, and testimonials. In terms of acknowledging the worth of information in guiding a rational and logical higher cognitive process, medicine has been at the forefront of adapting to modernity. However, physicians, nurses, and healthcare facilities are slow to embrace the most recent methods to completely use the information contained in EHRs.

There are many benefits of EHR which have been brought into healthcare. Some of them are listed below.

- Greater patient care
- Improved public health
- Ease of workflow
- Lower healthcare costs

2.2.5.1 Greater Patient care

EHR will mechanically alert the treating doctor to potential problems (such as allergies or intolerances to certain medicines). EHRs may be accessed from nearly any medical facility, which is very helpful for doctors assessing non-local patients.

2.2.5.2 Improved Public health

EHRs give valuable information to clinical researchers, serving to advance medical data and also the development of treatments for common health issues (such as viral outbreaks).

A standardized health IT system will give insights into how widespread an endemic is, enabling preventive measures (such as increased respiratory disorder shot production) to be put in place rather more quickly.

2.2.5.3 Ease of Workflow

Medical billers and coders are a number of the most-impacted allied physicians, and—according to the Bureau of Labor Statistics—demand for this sector is predicted to grow by 13% from 2016 to 2026. The introduction of EHRs has just made life easier for medical billers and coders.

Entering information into a processed system is way less time-consuming than paper-based strategies, and it reduces the danger of errors in patient information and money details. Accessing patient records digitally additionally permits medical committal-to-writing experts to work from home, increasing efficiency and productivity.

2.2.5.4 Lower Healthcare costs

According to recent research, the shift from paper-based patient records to electronic records reduces the costs of outpatient care by 3%.

2.2.5.5 Disadvantages of EHR

Theoretically, shifting to EHRs ought to change everything for the better. Sadly, there are some kinks that also need to be smoothed out. Instead of a records system that works fluidly, several networks lack interconnectivity, which suggests that several do not have the ability to speak to each other. Sometimes, this lack of communication will place patients' health at risk.

2.2.6 Big Data

Big data is the buzzword nowadays. It is seen everywhere, particularly within the healthcare business. Historically, the massive quantity of data generated by the healthcare business was held as a hard copy. This information has the potential to support a wide range of healthcare and medical functions. The conversion of such information is termed big data. All of the data that is associated with patient healthcare and wellbeing makes up big data.

The wide diversity of big data and also the pace at which it is managed makes it overwhelming. It includes clinical information from CPOE and clinical decision support systems, physicians' written notes and prescriptions, medical imaging, laboratory, pharmacy, insurance, and other administrative data; patient data in electronic patient records (EPRs); machine-generated or sensor data, such as from monitoring vital signs; social media posts, together with Twitter feeds, blogs, status updates on Facebook and different platforms, and web pages; and less patient-specific data, together with emergency care information, news feeds, and articles in medical journals.

Big data is extremely helpful within the healthcare business. Over the past decade, electronic health records (EHR) have been widely adopted in hospitals and clinics worldwide. Important clinical data and a deeper understanding of patient sickness patterns may be studied from such information. It will help to boost patient care and improve efficiency. Sometimes, this lack of communication will place patients' health at risk.

With its diversity of format, type, and context, it is tough to merge big healthcare data into standard databases, making it tremendously difficult to process, and hard for business leaders to harness its important promise to remodel the industry.

Despite these challenges, many new technological enhancements are permitting healthcare big data to be born again into helpful, weighted data. By leveraging applicable software package tools, big data is informing the movement toward value-based healthcare and is open to outstanding advancements, even while reducing prices. With the wealth of knowledge that healthcare data analytics provides, caregivers and administrators will currently make better medical and monetary decisions while still delivering an ever-increasing quality of patient care. But adoption of big data analysis in healthcare has lagged behind alternative industries thanks to challenges like privacy of health data, security, sliced knowledge, and budget constraints. Meanwhile, 80% of executives from financial services, insurance, media, entertainment, manufacturing, and supply firms surveyed report their investments in big data processing as "successful," and almost one in five declare their big data initiatives are "transformational" for his or her corporations.

There are at least two trends nowadays that encourage the healthcare business to embrace big data. The primary is the move from a pay-for-service model that

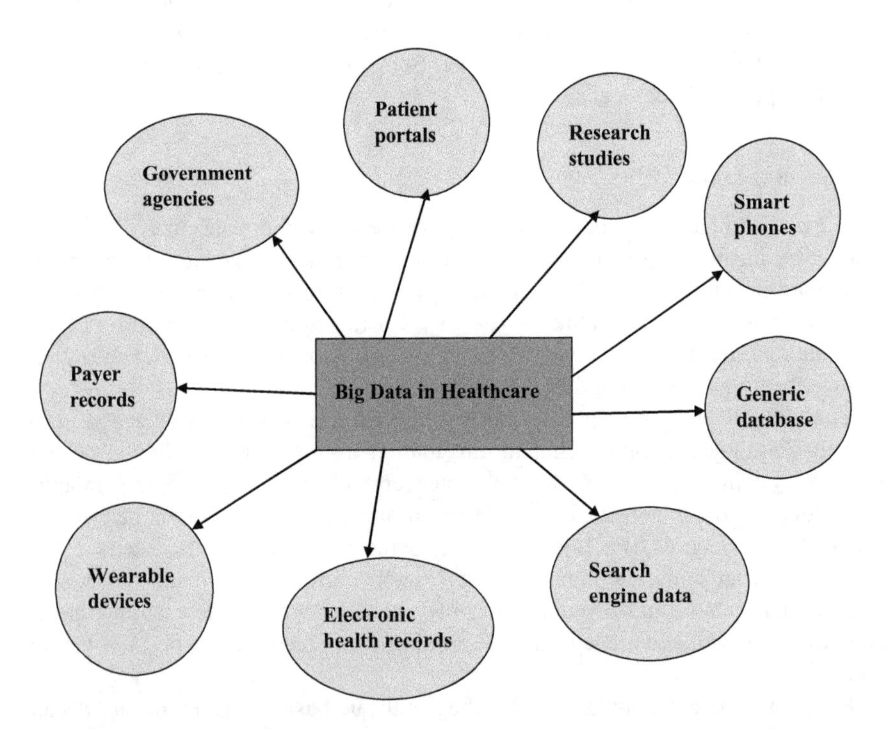

FIGURE 2.1 Sources of Big data.

financially rewards caregivers for performing procedures, to a value-based care model that rewards them for supporting the health of their patient populations. Healthcare data analytics can change the measuring and tracking of population health, thereby sanctioning this switch. The second trend involves exploiting big data analysis to deliver data that is evidence-based and can, over time, increase efficiencies and facilitate sharpening of our understanding of the most effective practices related to any disease, injury or ill health.

Undoubtedly, adopting the utilization of healthcare big data will remodel business, driving it from a fee-for-service model toward value-based care. In short, it will deliver on the promise of lowering healthcare prices while revealing ways in which to deliver superior patient experiences, treatments, and outcomes.

Using big data has multiple benefits such as

- Reducing healthcare costs
- Predicting epidemics
- Avoiding preventable deaths
- Improving quality of life
- Reducing healthcare waste
- Improving efficiency and quality of care
- Developing new drugs and treatments

2.2.6.1 Applications of Big Data in Healthcare

- Integrating big data with medical imaging.
- Telemedicine
- Reduce fraud and enhanced security
- Predictive analytics in healthcare
- Using health data for strategic planning
- Real-time alerting
- Electronic health records
- Enhancing patient engagement
- Smoother hospital administration
- Big data to fight cancer.

2.3 REMOTE HEALTH MONITORING

Remote health monitoring, also described as remote patient monitoring, is the technique of exploiting technology to observe patients in non-clinical environments, such as within the home. Once incorporated within the management of chronic diseases, remote health monitoring has the potential to considerably improve quality of life for patients and so it ought to come as no surprise that this technology is growing progressively standard. Remote health monitoring could be a specific field; however, associated technologies frequently share similar elements. Firstly, a monitoring device needs a sensor which might generate specific physiological information and wirelessly communicate this data to both the patient and aid professionals.

Shared information storage for this data is additionally crucial, both as software that may analyze health information and supply treatment recommendations and

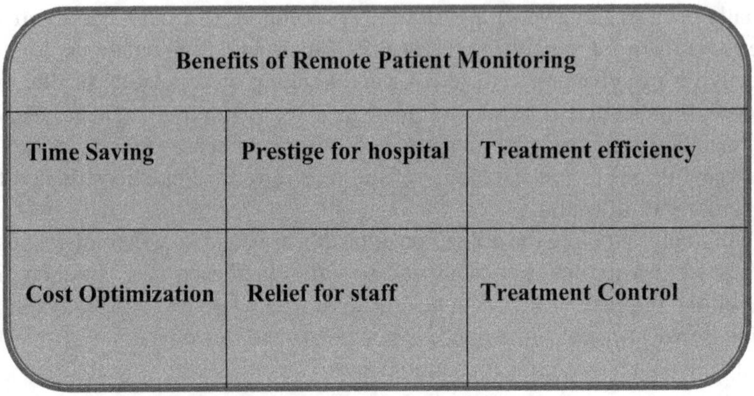

Benefits of Remote Patient Monitoring		
Time Saving	Prestige for hospital	Treatment efficiency
Cost Optimization	Relief for staff	Treatment Control

FIGURE 2.2 Benefits of Remote Health monitoring in Healthcare.

alerts. Health-monitoring technologies that accept smartphone apps have become progressively standard. These will aid patients with varied conditions; however, the technology is most generally used for monitoring heart conditions and diabetes. Diabetics need to regulate their weight, blood pressure and blood glucose levels to remain healthy.

2.3.1 Benefits of Remote Health Monitoring

- Improved education, support and feedback
- Improved quality of care
- Better access to care
- Daily assurance
- Cost-effective
- Makes healthcare services accessible
- Makes healthcare services efficient
- Makes healthcare consumer centric
- Improves patient's lifestyle
- Allows sending data from patients to doctors in real time.

Below is a figure which describes the key benefits of using remote health monitoring in healthcare units.

2.3.2 Challenges of Remote Health Monitoring

Some of the most general difficulties encountered with remote health monitoring in healthcare units are listed below.

- Confidentiality and privacy
- Users' attitudes
- Organizational and technological barriers
- Maintainance cost

- Implementation cost
- Costly modern systems
- Poor design and implementation
- Insufficient investment
- System incompatibility with personal tasks
- Data manipulation, rewriting, data misrepresentations
- Documentation mistakes
- Violation of patients' legal rights
- Decrease in face-to-face communication between doctors and patients
- Weight of wearable devices

Although the technology has several proven successes, its main flaws mask the fact that remote health monitoring will heavily count on patients taking a vigorous role in their own health and a few patients are more passive or forgetful than others. Wireless technologies also are not appropriate for some rural areas, and a few older patients might not know how to use modern technologies such as apps. Any collected health data additionally has to be encrypted and protected from hackers, and a few remote health-monitoring technologies are very expensive.

However, remote health monitoring will provide patients with additional power to keep an eye fixed on their health and may offer peace of mind for those managing chronic conditions. As technology will alert patients and healthcare professionals to slight physiological changes, any definitely dangerous conditions are also likely to be caught and treated earlier. The prices are high and therefore the technology still has to be refined; however, as those barriers bit by bit are countermined, remote monitoring is probably going to become a core part of preventive care in the longer term.

2.3.3 Obstacles of Remote Health Monitoring Usage

There are several obstacles in the path of Remote Health Monitoring (RHM) usage. Below are some of the points which describe these.

2.3.3.1 Is Not Accessible for Everyone

RHM needs good broadband connectivity that is difficult to attain for small healthcare establishments and rural hospitals. On the other hand, it is necessary to keep under consideration that not everybody owns a smartphone, and older people typically face difficulties in using modern gadgets, such as mobile phones.

2.3.3.2 Patients' and Doctors' Skepticism

RHM seems to be the least effective patient engagement initiative according to the NEJM Catalyst Insights Council survey. The researchers justify these statistics by the very fact that the employment of wearable remains unobtainable for everyone. By the way, doctors who took part in the survey did not notice any improvements in chronic disease management. Additionally, healthcare professionals added their expertise to express doubt that technology alone is probably going to alter the behavior of higher-risk patients. Doctors are involved in the difficulties they will face handling received information. Some patients are afraid that their personal health information will be obtained by third parties and used for dubious purposes.

2.3.3.3 The Necessity of Extra Custom Healthcare Software

Once the information is collected IT departments need to direct it from RHM devices to electronic medical history systems (EMRs) by means of multiple third-party applications.

2.3.3.4 Uncertain Reliability

Fashionable fitness wearables supporting physical activity are perceived to have giant variations of accuracy with error margins up to 25 percent. The reliability of RHM information is called into question also. For example, a review in JAMA dermatology showed that smartphone apps for skin cancer detection have a 30 percent failure rate. The lack of reliability is the most significant issue that has to be fixed before devices and applications might be utilized by healthcare suppliers.

RHM is extremely keen on the individual's motivation to manage their health. If it is not the patient's temperament to be a lively participant in their care, RHM implementation can probably fail. Price is additionally a barrier to its widespread use. There's an absence of compensation pointers for Remote Patient Monitoring (RPM) services, which can deter its incorporation into clinical observation. The shifting of untrustworthiness identified with RHM raises liability problems. There are no clear pointers of relevance as to whether or not clinicians need to intervene when they receive an alert in spite of the urgency. The continual flow of patient information needs a frenzied team of healthcare suppliers to handle the data, which may, in fact, increase the work. Though technology is introduced with the aim of extending potency, it will become a barrier to some healthcare suppliers that do not seem to be technological. There are common obstacles that health information processing technologies encounter that apply to RHM. Looking at the comorbidities monitored, RHM involves a wide choice of devices in its implementation. Standardization is needed for information exchange and ability among multiple elements. Moreover, RHM development is extremely keen on an intensive wireless telecommunications infrastructure, which cannot be on the market or possible in rural areas. Since RHM involves the transmission of sensitive patient information across telecommunication networks, data security could be a concern.

2.4 DISADVANTAGES OF USING IOT IN HEALTHCARE

2.4.1 Expensive

A progressively refined health technology does not return low cost. We have got to grasp that each first-world national healthcare system faces a variety of challenges; one among those is the aging population. People live longer. This suggests an increased health need; however, the working population generating financial gain to procure the healthcare system is reduced. Therefore one thought would be: Is the high price that comes with technology economically viable for the government?

2.4.2 Time-consuming Adaption

As we know, technology is continually evolving. Many times there will be new software systems, new upgrades, and a brand new way of doing things. In order to keep a competitive edge, hospital employees have got to sustain such changes. This will be a struggle for a few, particularly for the older workers.

2.4.3 Technology Dependent

Once the workers have adapted to the new manner of labor, there comes a further drawback. It is not uncommon for a computer system to face technical errors. The healthcare informatics system is no exception. This drawback is particularly crucial within the Accident & Emergency (A&E) Department. Varied departments within the hospital are interconnected by a standard data system. Once one department is down, others are affected. For instance, a patient is rushed into the A&E Department. Once there is a mistake retrieving blood analysis data, the remainder of the procedures following it will be delayed. This can cause great inconvenience or worse; it is even going to have adverse effects on the patient's health condition.

2.4.4 Susceptibility to Network Hackers

Patients' medical history and data should be kept confidential for legal reasons. But even if the healthcare system network is equipped with security measures it is possible for network hackers to extract this information, which is now a matter of concern for Health Informatics.

2.4.5 Technical Requirements

For an IoT network to give value to a business, it should work as one, cohesive system. From a technical perspective, the truth is that IoT is commonly fragmented and lacks ability. To combat this, platforms should be ready to operate across devices no matter the build, manufacturer or business. Overcoming compatibility problems may be a

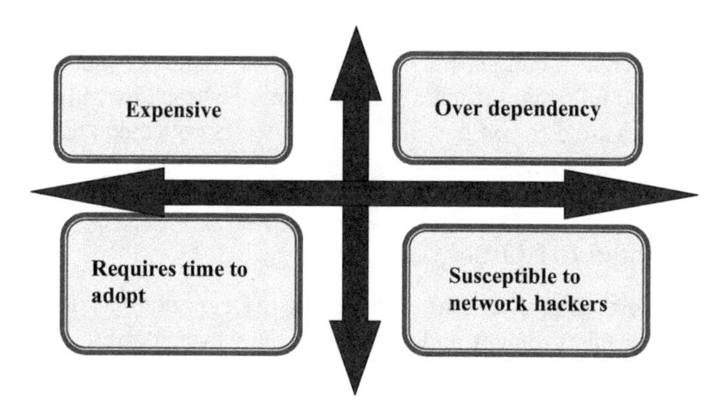

FIGURE 2.3 Disadvantages of IoT in Healthcare.

vital IoT hurdle; however, more businesses are beginning to modify increased ability through open-source development.

2.4.6 Regulatory Concerns

IoT implementations are collecting massive amounts of information that would probably be sensitive or harmful if exposed. This includes personal knowledge concerning workers or customers, as well as proprietary business knowledge concerning operations and internal processes. From a restrictive perspective, privacy considerations, such as clarifying who will access IoT knowledge and the way that data is employed, should be addressed. Governments and business bodies have to be compelled to set standards and rules for the varied industries to make sure that knowledge is not misused.

2.4.7 Scalability

Businesses typically successfully develop an IoT application with several devices in a single location. However, they later discover that measurability is a problem. It is crucial that the planning of a system caters for extra resources.

2.4.8 Time Constraints

The rolling out of IoT is an extended and expensive endeavor for businesses. Speedy changes in technology mean that businesses run the danger of any new technology system becoming obsolete as soon as it is installed. To profit from the advantages of a brand new IoT system, businesses should attempt to eliminate as many of the obstacles as possible from the company's development method to achieve a quick and economical rollout. Equally they have to make sure that they partner with competent service suppliers that may meet their technology wants and modify agile solutions.

2.4.9 Security

IoT devices are typically susceptible to security breaches due to poor design. This may have a significant impact on company knowledge security, and also put pricey IoT-related equipment in danger. For this reason, IoT requires robust authentication methods, encrypted information and a platform that may track irregularities on a network. If businesses are clear on how IoT information is collected and used, shopper confidence in IoT will grow.

2.4.10 Managing IoT Obstacles

The inflow of information as a result of adopting IoT can modify businesses to considerably improve management and operations. However, despite the prospect of IoT remodeling businesses, implementing associated IoT resolutions will be fraught with complexities. Whereas deploying IoT technologies has huge potential, in order to exploit these advantages IoT challenges ought to be addressed effectively, and potential challenges and security problems ought to be overcome to ensure IoT success. Problems such as technology, regulation, measurability, roll out and security all need to be addressed.

2.5 CHALLENGES WITH MANAGING IOT TECHNOLOGIES

While the IoT can bring important advantages, they will be difficult to implement. Forbes Insights recently surveyed almost five hundred executives and, when asked about their greatest challenge in building their IoT capabilities, twenty-nine said it had been the standard of IoT technology. This is not stunning. In some cases, IoT platforms must support thousands of vendors, dozens of standards, and be able to scale to countless devices, along with creating and receiving billions of messages.

IoT-based solutions are generally created from a number of technologies, some already existing and a few entirely new. Everything has its own path of development, and once they are combined, they will produce an atmosphere that is complicated and speedily dynamical. Here are four challenges with managing IoT technologies these days.

2.5.1 Integrating New Technologies into Existing Environments

In the era of the smartphone, it will appear as if each machine is connected and sharing data; however, that is not the case. Within the client world, a combination of technologies competes for dominance, and standardization remains elusive. As a result, comparatively few homes, appliances, or other commodities are literally IoT-enabled and connected. In the industrial world, it gets even more sophisticated owing to the character of the investments. Capital instrumentality that has been within the field for twenty years or so is not invariably a viable target for replacement, as a stove or refrigerator could also be within the shopper world. Retrofitting is usually the sole realistic resolution to bring IoT capabilities to existing instrumentation. However, retrofitting is neither easy nor assured. Whereas connecting legacy equipment and systems offers huge advantages and is a very important step within the IoT initiatives at several industrial corporations, the hurdles to implementation may be formidable.

That said, corporations are creating vital strides within this space. They are adding complete sensors and cameras to existing environments and devices to monitor and collect information regarding machine performance and health. These sensors attached to existing devices and connect with gateways to firmly collect and transmit information, which might then be analyzed.

2.5.2 Managing Protocol Complexity

Another huge challenge within the development of the IoT is the immense variety of protocols. Some of the common standards include:

- BLE (Bluetooth low energy)
- ZigBee
- Z-Wave

- Thread
- We-Mo

In some ways, BLE, ZigBee, Z-Wave, and Thread are similar. They are all wireless technologies that use mesh networks to wirelessly connect and network IoT devices while not involving a cellular or Wi-Fi signal. However, they differ in the frequency they use, and they vary in operation and therefore the variety of devices they will support at a given time. We-Mo, however, needs Wi-Fi, which eliminates the requirement for a hub or controller, and permits devices to attach directly via the net. Two of the massive disadvantages of this technique are that it needs a lot more power and processing capability than alternative, lower-energy choices.

Again, this can only be a brief list; the quantity of protocols is in depth. Each has its advantages and downsides; however, since there is no single common standard, businesses should confirm the correct protocol for every use case and make sure the technologies they select are compatible with their overall platform. As standards are still evolving, it is going to be advantageous to exchange or upgrade the method.

2.5.3 Networking Challenges

Beyond the various completely different protocols and disparate hardware, there are basic networking challenges that have got to be addressed to make IoT-enabled devices a reality. The primary step is to make sure that information is flowing quickly and reliably. Security is additionally crucial, as IoT devices are often evolving into targets for hackers and cyber terrorists. Once devices connect, they have to certify, information should be encrypted, and they ought to communicate their presence and activity.

Power consumption and bandwidth present alternative distinctive challenges. During a scenario where thousands of devices are communicating with each other, frequent communication and transmission may be a drain on battery-operated devices. In those cases, minimal, economical power usage is vital. During a scenario where thousands of devices are communicating over wireless networks, bandwidth will become a priority, and costs will increase quickly. The goal should be to keep IoT information streams as compact and economical as possible.

2.5.4 Best Practices in the Era of IoT

Within the IT world, best practices are generally described as procedures that are well known and considered to be the most effective. Nowadays there's an absence of best practice to help businesses write code, manage the life cycle of certain IoT-related hardware and software systems, and address the distinctive kinds of breaches that may occur, together with intrusions that are initiated at the device level. Without best practices as a road map, programmers and IT professionals are traveling in unmapped waters. Consider the Mirai botnet attack in October 2016. Throughout this incident, IT professionals saw firsthand how prolific a breach can be. Although the incident was damaging, many things were learned, together with the importance of getting an IoT security strategy and the concept of fast response.

As the IoT continues to proliferate, there are guaranteed to be growing pains. Hardware can still advance and improve. The software system can become a lot more refined. New standards, protocols and connectivity choices can become much more prevalent. However, businesses should ensure that their new capabilities stay compatible with legacy systems. With this sort of approach, businesses will simply handle the speed of change that comes with the IoT and very much notice its advantages.

2.6 SECURITY THREATS OF IOT

The Internet of Things (IoT) delivers substantial advantages to end-users. However, it also brings new security challenges. Part of the central security issue is that connected devices share implicit trust. This shared trust between connected devices means the devices automatically transmit their information to every alternative straight away upon recognition while not initially running any malware detection tests. The worst-case eventualities of those IoT security dangers lead to physical harm or maybe the loss of life.

Connected devices are creating pleasant experiences for consumers; however, they also represent current targets for hackers. The Internet of Things (IoT) and cyber-criminal activity share two vital traits: they are mostly invisible to the eye, and they surround us at any given moment.

As additional organizations use a combination of sensors and complicated software system applications to create smart homes, smart workplace environments, and even smart cities, the results typically feel magic. Lights turn on after you enter a room. A piece of machinery proactively requests an upgrade to stop breaking down. A retail store automatically restocks a shelf before customers become annoyed over missing things. These are all ways in which the IoT makes technology omnipresent and seamless. Unfortunately, the foremost prospering cybercriminals behave in an almost identical way. Hacking databases, attacking websites, and stealing passwords seldom involves a face-to-face encounter. Once technology becomes essential, the security problems associated with the technology tend to mount. Over time, these problems have transitioned from email to text messages, from desktop PCs to smartphone and currently to the IoT.

The Internet of Things (IoT) may be a quickly growing phase of the internet. Whereas different parts of the internet are dependent on individuals exchanging data, IoT allows devices to gather data, transmit data and receive data. It is easier to think about IoT as similar to online, email or social networks; however, rather than connecting individuals, it connects smart machines.

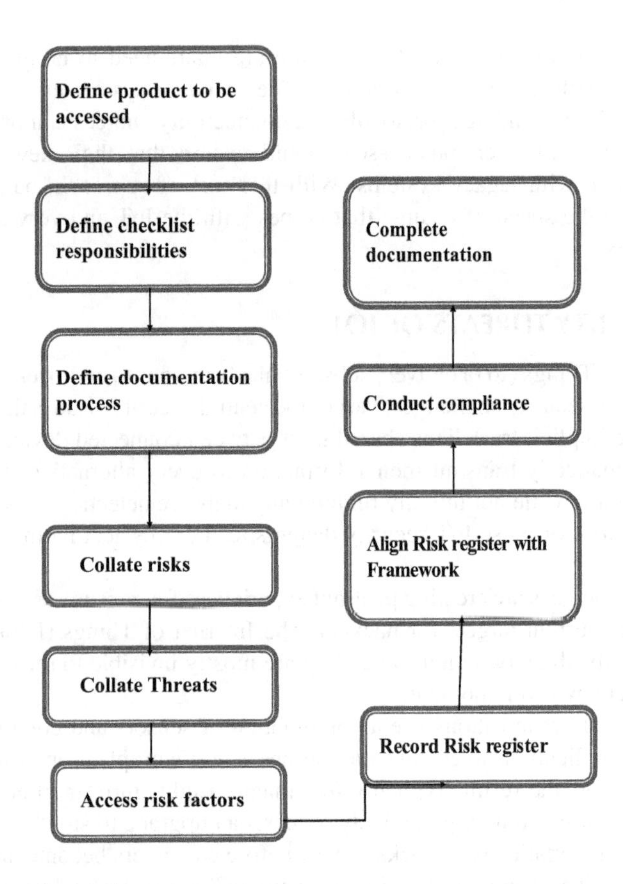

2.6.1 Vulnerability

The most basic and easy-to-pick threat to IoT devices is their vulnerability. Corporations providing IoT solutions begin with addressing this issue initially before looking at the underlying software system. We are also compelled to perceive that vulnerability is often of two types: hardware and software. Hardware vulnerability is commonly hard to discover or penetrate. However, it is even harder to repair or overhaul the injury. Software vulnerability points toward a poorly written algorithmic program or a line of code with a backdoor. This backdoor will simply give access to intruders prying for such opportunities.

2.6.2 Easy Exposure

This can be one of the most basic problems faced by the IoT business. Any device, if unattended or exposed to troublemakers, is an open invitation to discomfort. In most cases, IoT devices are not flexible to third-party exposure; they either expose or are simply accessible to anyone.

This means that an attacker will either simply steal the device, connect the device to a different device containing harmful information, or attempt to extract cryptological secrets, modifying the programming or perhaps substituting those devices with malicious ones over which the intruder has complete management.

2.6.3 Threats

Threats are often of two types: an individual's threat or a natural threat. Any threat arising from natural occurrences like earthquakes, hurricanes, floods, or fires will cause severe injury to IoT devices. We regularly take a backup or produce contingency plans to safeguard information. However, any injury caused to the devices physically cannot be repaired.

Nowadays, IoT solutions have matured over time. Devices, today, have evolved to be waterproof. It will be a long journey before IoT solution suppliers come up with something that is fireproof or earthquake-proof. On the contrary, we tend to do everything in our power to curb any human threats to IoT devices. These threats are typically malicious attacks.

2.6.4 Insecure Web Interface

The security issues relating to the Internet of Things are that the built-in IoT allows the end-user to interface with devices while at the same an attacker or intruder may also gain an unauthorized access to these devices.

Some of the security issues are listed below.

- Account enumeration
- Weak default credentials
- Credentials exposed in network traffic
- Cross-site scripting (XSS)
- SQL-injection
- Session management
- Weak account lockout settings.

The solution to the above problems is

- Default passwords and default usernames must be changed during initial setup.
- Ensuring password-recovery mechanisms are robust and do not supply an attacker with information indicating a valid account.
- Ensure that web interface is not susceptible to XSS, SQLi or CSRF.
- Ensure that credentials are not exposed in internal or external network traffic.
- Weak passwords should not be allowed.
- Account lockout after 3–5 failed login attempts.

2.6.5 Insufficient Authentications

This area deals with mechanisms insufficient to authenticate IoT devices. These give rise to such issues as

- Lack of password complexity.
- Poorly protected credentials.
- Lack of two-factor authentication.
- Insecure password recovery.
- Privilege escalation.
- Lack of role-based access control.

The solution to the above problems is

- Ensure that strong passwords are required.
- Ensuring granular access control is in place when necessary.
- Ensuring credentials are properly protected.
- Implement two-factor authentications where possible.
- Ensuring those password-recovery mechanisms are secure.
- Ensuring re-authentication is required for sensitive features.
- Ensuring options are available for configuring password controls.

2.6.6 Insecure Network Devices

Here are some of the problems related to the insecure network devices with IoT:

- Vulnerable services
- Buffer overflow
- Open ports via UPnP
- Exploitable UDP services
- Denial-of-service
- DoS via network device fuzzing.

The solution to the above problems is

- Ensuring only necessary ports are exposed and available. Ensuring services are not vulnerable to buffer overflow and fuzzing attacks.
- Ensuring services are not vulnerable to DoS attacks which can affect the device itself or other devices and/or users on the local network or other networks.
- Ensuring network ports or services are not exposed to the internet via UPnP for example.

2.6.7 Lack of Transport Encryption

This area deals with the data or information being exchanged in an unencrypted format.

- Unencrypted services via the internet.
- Unencrypted services via the local network.
- Poorly implemented SSL/TLS.
- Misconfigured SSL/TLS.

The solution to the above problems is

- Ensuring data is encrypted using protocols such as SSL and TLS while transiting networks.
- Ensuring other industry-standard encryption techniques are utilized to protect data during transport if SSL or TLS are not available.
- Ensuring only accepted encryption standards are used and avoid using proprietary encryption protocols.

2.6.8 Privacy Concerns

Collection of unnecessary personal information is the main concern of patients. The information of patients should be legal and it should not be accessed by any third party.

Some points should be kept in mind while dealing with privacy threats. They are listed below:

- Ensuring only data critical to the functionality of the device is collected.
- Ensuring that any data collected is of a less sensitive nature.
- Ensuring that any data collected is de-identified or anonymized.
- Ensuring any data collected is properly protected with encryption.
- Ensuring the device and all of its components properly protect personal information.
- Ensuring only authorized individuals have access to collected personal information.
- Ensuring that retention limits are set for collected data.
- Ensuring that end-users are provided with "Notice and Choice" if data collected is more than what would be expected from the product.

2.6.9 Insecure Cloud Interface

This point concerns security issues related to the cloud interface used to interact with the IoT devices.

- Account enumeration.
- No account lockout.
- Credentials exposed in network traffic.

The solution to the above problem is

- Default passwords, default usernames to be changed during initial setup.
- Ensuring user accounts cannot be enumerated using functionality such as password reset mechanisms.
- Ensuring account lockout after 3–5 failed login attempts.
- Ensuring the cloud-based web interface is not susceptible to XSS, SQLi or CSRF.

FIGURE 2.4 Demonstration of Web security.

- Ensuring credentials are not exposed over the internet.
- Implement two-factor authentications if possible.

2.7 CONCLUSIONS

This chapter discussed the various types of application of the Internet of Things (IoT) in the healthcare industry. We have focused on the various types of application of IoT in the healthcare industry; the major security issues and the challenges present in the way of successful implementation of IoT in the industry. IoT is often blamed for delusions of grandeur and therefore the conviction that it is a big deal. The fact is that it is a giant deal and destined to grow larger. IoT is certainly a giant deal and it is only planning to get larger with the passage of time. Sadly, the larger it gets the greater a target is on its back. Likewise, all the related threats and IoT trends can get larger. Manufacturers and others linked with the IoT trade may have to be compelled to get serious regarding protection problems and threats.

REFERENCES

1. Alok Kulkarni, Sampada Sathe. "Healthcare applications of the Internet of Things: A Review". *(IJCSIT) International Journal of Computer Science and Information Technologies*, 5 (5), 2014, 6229–6232.
2. B. Sobhan Babu, K. Srikanth, T. Ramanjaneyulu, I. Lakshmi Narayana. "IoT for Healthcare" *International Journal of Science and Research (IJSR)* 2014.
3. Shubham Banka, Isha Madan, S.S. Saranya. "Smart Healthcare Monitoring using IoT. *International Journal of Applied Engineering Research ISSN 0973-4562*, 13 (15), 2018, 11984–11989.

4. Ananda Mohon Ghosh, Debashish Halder, SK Alamgir Hossain "Remote Health Monitoring System through IoT" 5th International Conference on Informatics, Electronics and Vision (ICIEV)
5. Damian Dziak,*, Bartosz Jachimczyk, Wlodek J. Kulesza "IoT-Based Information System for Healthcare Application` Design Methodology Approach" *Appl. Sci.* 2017, 7, 596; doi:10.3390/app7060596 www.mdpi.com/journal/applsci
6. Harold Thimbleby. "Technology and the future of healthcare." [*Journal of Public Health Research*, 2013; 2:e28]
7. Burgos S. Medical information technologies can increase quality and reduce costs. *Clinics*, 2013, 68(3):425, http://dx.doi.org/10.6061/clinics/ 2013(03)LE04.**Jo**
8. https://dzone.com/articles/the-biggest-security-threats-and-challenges-for-io Health Research 2013; vo
9. www.iotforall.com/7-most-common-iot-security-threats-2019/28
10. www.ubuntupit.com/25-most-common-iot-security-threats-in-an-increasingly-connected-world/
11. www.sdxcentral.com/5g/iot/definitions/iot-security/

3 "IoT" *Bright Future in Healthcare Industry*

[1]Subhashree Sahoo, [2]Debabrata Dansana, and [2]Raghvendra Kumar
[1]Pondicherry University, Kalapet, Pondicherry, India
[2]GIET University, Gunpur, India

CONTENTS

3.1 INTRODUCTION

The Web of Things or IoT is a domain of related physical things that are accessible through the internet. The 'thing' in IoT could be a person with a heart screen or a vehicle with characteristic sensors, for instance, protests that have been named an IP address and can accumulate and trade data over a framework without manual assistance or mediation. The implanted innovation in the items makes them interface with internal states or the external condition, which accordingly impacts the choices taken. The importance of the Internet of Things has progressed in light of the mixing of various advances, constant examination, AI, item sensors, and installed frameworks. Conventional fields of implanted frameworks, remote sensor frameworks, control systems, motorization, and others all add to enabling the Internet of Things. In the client showcase, IoT advancement is most synonymous with things identifying with the possibility of the "smart home", covering contraptions and devices, such as lighting establishments, indoor controllers, home security frameworks and cameras, and other home machines that help in any event one's customary conditions, and can be constrained by methods for devices related with that condition, such as smartphone and smart speakers. The IoT concept has faced prominent examination, especially regarding assurance and security concerns related to these devices and their desire for unavoidable existence.

The possibility of an arrangement of smart gadgets was discussed as early as 1982, with a changed Coke candy at Carnegie Mellon University transforming into the principle internet-related mechanical assembly, prepared to report its stock and whether as of late stacked drinks were cold or not. Weiser's 1991 paper on ubiquitous figuring, "The Computer of the 21st Century", similarly as academic settings, for instance, UbiComp and PerCom made the contemporary vision of the IoT. In 1994, Reza Raji delineated the thought in IEEE Spectrum as meager groups of data to a broad game plan of center points, to join and motorize everything from home machines to entire mechanical offices". Somewhere between 1993 and 1997, a couple of associations proposed game plans like Microsoft's at Work or Novell's NEST. The field grabbed vitality when Bill Joy envisioned contraption to-device correspondence as a part of his "Six Webs" structure, shown at the World Economic Forum at Davos in 1999. The expression "Web of Things" was likely begat by Kevin Ashton of Procter and Gamble, later MIT's Auto-ID Center, in 1999, be that as it may, he slants toward the articulation "Web of Things". By that point, he saw Radio-repeat ID (RFID) as fundamental to the Internet of Things, which would empower PCs to manage each individual thing. An examination article referencing the Internet of Things was submitted to the social occasion for Nordic Researchers in Norway, in June 2002, which was preceded by an article dispersed in Finnish in January 2002. The use delineated there was made by Kary Främling and his group at Helsinki University of Technology and even more eagerly organizes the propelled one, for instance, an information system establishment for executing smart, related items.

3.2 SCOPE OF IOT

IoT embedded gadgets enable consumers to attain deeper understanding, examination, and integration within the gadgets. These gadgets are designed accordingly to improve the functionality of those diverse areas simultaneously increasing their accuracy. IoT uses custom technology that was already present and continuous improving technology for identifying, networking of devices, and artificial intelligence. IoT comprises continuous advances in the software system, falling equipment expenses, and in vogue demeanors toward innovation. It is a new and propelled segments bring real changes inside the conveyance of product, merchandise, and benefits; and furthermore the social, financial, and political effect of these advances. The foremost necessary features of IoT embrace AI, property, sensors, viable engagement, and miniature gadgets use.

- **Artificial intelligence & IoT** – IoT makes partially something "smart" that means it upgrades every part of life with the ability to use information assortment, AI algorithms, and networks. We can say one thing is much easier to notice by an individual that when his necessity things like milk and cereals is going to finish by just looking into the refrigerator and cabinets and to then place an order along with most well-liked grocery shops.
- **Connectivity & IoT**– Previous endorsing innovation that is utilized for systems administration, and uncommonly utilized for IoT organizing, that these systems are no longer exclusively attached to real providers. Systems can exist on a way littler and less expensive scale though as yet being reasonable. IoT makes these little systems between its framework gadgets.
- **Sensors & IoT** – IoT loses its refinement while not sensors. They go about as characterizing instruments which change IoT from a standard detached system of gadgets into an incredible framework able to do true mix.
- **Helps in Active Engagement** – It can be a great deal of the present cooperation with associated innovation occurs through uninvolved commitment. IoT

presents a fresh out of the plastic new worldview for dynamic substance, item, or administration commitment.

- **Miniature Devices** – Devices, as anticipated, decreased, less expensive, and all the more dominant after some time. IoT endeavors reason made little gadgets to convey its exactness, adaptability, and flexibility.

3.2.1 Information Accumulation

This product framework oversees detecting, estimations, lightweight data sifting, lightweight data security, and a total of data. It uses bound conventions to help sensors in interfacing with a period, machine-to-machine systems. At that point, it gathers data from numerous gadgets and conveys it as per settings. It conjointly works backward by dispersing data over gadgets. The framework, in the end, transmits all gathered data to a local server.

3.2.2 Device Integration

Programming supporting reconciliation ties all framework gadgets to make the body of the IoT framework. It guarantees obligatory participation and stable systems administration between gadgets. These applications are the molding programming framework innovation of the IoT organizes because of, while not them, it is anything but an IoT framework. They deal with the changed applications, conventions, and confinements of each gadget to allow correspondence.

3.2.3 Real-Time Analytics

These applications take data or contribution from various gadgets and convert them into reasonable activities or clear examples for human examination. They dissect information bolstered various settings and plans in order to perform mechanization related errands or offer the data required by the business.

3.2.4 Apps and Method Abridgment

These applications expand the scope of existing frameworks and programming framework to allow a more extensive, increasingly successful framework. They coordinate predefined gadgets for explicit capacities like allowing certain cell phones or designing instruments get to. It supports improved efficiency and extra exact data arrangement.

3.3 HEALTHCARE INDUSTRY

The healthcare business is in a surpassing condition of decent depression. These services are costlier than any time in recent memory, the overall populace is maturing and in this manner, the assortment of ceaseless sicknesses is on expansion. What we tend to are drawing closer could be a world any place fundamental medicinal services would wind up distant to the larger part, a curiously large segment of society would go useless as a result of adulthood and people would be a ton of defenseless to endless affliction. Is it accurate to say that it isn't the highest point of earth we tend to suspected? Regardless of IoT application improvement is at your salvage. While innovation cannot prevent the populace from maturing or annihilate interminable ailments legitimately, it will at least form human services simpler on a pocket and in term of openness. Therapeutic demonstrative devours a larger than average a piece of clinic bills. Innovation will move the schedules of medicinal checks from a clinic (emergency clinic driven) to the patient's (home-driven). The correct diagnosing likewise will diminish the need for hospitalization. Another worldview, called the Internet of Things (IoT), has inside and out pertinence in changed territories, together with medicinal services. The total use of this worldview inside the medicinal services space could be a common expectation because it licenses therapeutic focuses to play out a great deal of relevantly and patients to get higher treatment. With the usage of this innovation based medicinal services procedure, there are unique points of interest that may improve the standard and strength of medications and thus improve the wellbeing of the patients.[1]

In the previous decade, web associated gadgets are acquainted with patients in various structures. Regardless of whether information originates from vertebrate screens, Electro-cardiograms, temperature-screens or glucose levels, interest well-being information is huge for a couple of patients, albeit a few of those measures required line up communication with a medicinal service gifted. However, the use of IoT gadgets has been instrumental in conveying extra important, constant information to specialists and changes the need for direct patient-doctor collaboration. Early utilization of IoT in medicinal services conjointly encased smart beds that notice once they are involved and once a patient is attempting to get up. A smart bed may likewise change itself to ensure adequate weight and backing is connected to the patient while not the manual cooperation of medical attendants. Another zone any place smart innovation rapidly wound up quality in social insurance is once including home prescription distributors. These containers precisely move data to the cloud once the medicine is not taken, or different markers that the consideration group should be cautioned.[2]

3.3.1 Benefits of IoT in Healthcare Industry

3.3.1.1 Information Agglomeration and Examination

A huge amount of data that a medicinal services gadget sends during the exceptionally brief time because of their constant application is hard to store and oversee if the entrance to the cloud is inaccessible. Notwithstanding for medicinal services providers to aggregate data beginning from numerous gadgets and sources and break down it physically could be a vigorous wager. IoT gadgets will gather report and investigations the information progressively and slice the prerequisite to store the information. This all will happen to overcloud with the providers exclusively acquiring access to conclusive reports with charts. Also, consideration activities grant associations to prompt significant social insurance investigation and information-driven bits of knowledge that accelerate basic leadership and are a little sum powerless against blunders.[2]

3.3.1.2 Following and Cautions

The on-time alarm is pivotal inside the occasion of risky conditions. IoT licenses gadgets to collect significant information and move that information to specialists for the ongoing interest though dropping warnings to people concerning essential components by means of portable applications and option coupled gadgets. Reports and alarms give a firm supposition two or three patients' condition, regardless of place and time. It conjointly enables work to knowledgeable determinations and supply on-time treatment. Hence, IoT allows constant cautioning, following, and recognition, which permits dynamic medicines, higher precision, able intercession by specialists and improves total patient social insurance conveyance results.[7]

3.3.1.3 Remote Medical Assistance

In case of a crisis, patients will contact a specialist who is a few kilometers away with smart versatile applications. With quality arrangements carelessness, doctors will in a split second check the patients and set up the diseases in a hurry.

Additionally, different social insurance conveyance fastens that are forecast to make machines that may appropriate drug on the possibility of the patient's medicine and illness related learning available by means of coupled gadgets. IoT can improve the patient's consideration inside the clinic. This, therefore, can cut on individuals' region on medicinal services.[2]

3.3.1.4 Simultaneous Reporting and Monitoring

Real-time observing by means of associated gadgets will spare lives inside the occasion of a medicinal crisis like coronary illness, diabetes, respiratory disease assaults, and so forth. With constant recognition of the condition set up by proposes that of a smart therapeutic gadget associated with a smartphone application, associated gadgets will gather restorative and option required wellbeing information and utilize the information relationship of the smartphone to move gathered data to a medicinal professional. The IoT gadget gathers and moves wellbeing information: circulatory strain, oxygen, and glucose levels, weight, and ECGs. This data is kept inside the cloud and might be imparted to an authorized individual, who might be a therapeutic specialist, your protection guarantor, a working together wellbeing firm or partner outer counsel, to allow them to appear at the gathered data regardless of their place, time, or gadget.[8]

3.3.1.5 Start to Finish Availability and Moderateness

IoT will automate patient consideration advance with the help of social insurance quality arrangement and option new innovations, and cutting-edge human services offices. IoT permits capacity, machine-to-machine correspondence, information trade, and data development that produce help administration conveyance powerful. Availability conventions: Bluetooth lupus, Wi-Fi, Z-wave, ZigBee, and elective present-day conventions, medicinal services workforce can adjustment the strategy they spot affliction and infirmities in patients and may conjointly start progressive manners by which of treatment. Subsequently, innovation-driven arrangement cuts down the worth, by diminishing unessential visits, using higher quality assets, and raising the portion and planning.[10]

3.3.1.6 IoT and Patients

Gadgets inside the kind of wearable like wellness groups and option remotely associated gadgets like circulatory strain and heartbeat rate watching sleeves, glucometer, and so forth offer patients access to customized consideration. These gadgets are tuned to provoke calorie tally, practice check, arrangements, circulatory strain varieties, and far extra. IoT has changed individuals' lives, especially more seasoned patients, by endorsing a consistent quest for wellbeing conditions. This joins a noteworthy effect on people living alone and their families. On any unsettling influence or changes inside the standard exercises of an individual, a ready component sends the sign to individuals from the family and restless wellbeing providers.

3.3.1.7 IoT and Physicians

Victimization wearable and elective home watching instrumentality implanted with IoT, doctors will monitor patients' wellbeing extra adequately. They will follow patients' adherence to treatment plans or requirement for quick medicinal consideration. IoT permits human services experts to be extra vigilant and associate with the patients proactively. Data gathered from IoT gadgets will encourage doctors to set up the least complex treatment technique for patients and achieve normal results.

3.3.1.8 IoT and Hospitals

Other than watching patients' wellbeing, there are a few elective zones any place IoT gadgets are extremely useful in emergency clinics. IoT gadgets marked with sensors are utilized for interest the ongoing area of medicinal instrumentality such as wheelchairs, defibrillators, nebulizers, component siphons, and elective watching instrumentality. Perusing of the medicinal specialists at totally various areas can even be dissected continuously. The spread of diseases might be a noteworthy worry for patients in emergency clinics. IoT empowered cleanliness watching gadgets encourage in keeping patients from getting contaminated. IoT gadgets moreover encourage in quality administration like drug store interior control, and natural viewing, for instance, checking icebox temperature, and mugginess and temperature the executives.

3.3.1.9 IoT for Insurance Corporations

There are different open doors for wellbeing safety-net providers with IoT-associated wise gadgets. Protection partnerships will use data caught through wellbeing watching gadgets for his or her endorsing and claims activities. This data can alter them to watch misrepresentation claims and set up prospects for guaranteeing. IoT gadgets bring straightforwardness among back up plans and clients inside the guaranteeing, evaluating, cases to deal with, and hazard appraisal forms. Inside the light of IoT-caught information-driven determinations in all task forms, clients can have sufficient permeability into the basic idea behind each call made and strategy results. Guarantors may supply motivations to their clients for exploitation and sharing wellbeing information created by IoT gadgets. They will compensate clients for victimization IoT gadgets to remain track of their standard exercises and adherence to treatment plans and precaution wellbeing measures. This can encourage safety-net providers to reduce asserts significantly. IoT gadgets can even adjust protection enterprises to approve asserts through the information caught by these devices.[7]

3.3.1.10 Redefining Healthcare

As the IoT specific products used in the healthcare industry a huge amount of data is generated. It is necessary to hold the potential of the data and transform it into healthcare. In the below figure we have described the architecture of IoT which is a four-stage architecture, and basically, these stages are of processes. All these four stages are interconnected together in such a way that data used in one stage yields its value in the next stage.[2]

Step-1

The first-step consists of numerous interconnected devices. They are used to collect information from various fields. These devices include a sensor, actuator, monitors, detectors, camera systems, etc.

Step-2

In the second step, the data received in the sensor and devices are cannot be used directly in further for data processing because the data are available is in analog form, so the data must be transformed and aggregated to digital-form for processing.

Step 3

In the third step, data is pre-processed, standardization of data is done and moved to the data-cloud.

Step-4

In the fourth step, data analysis operation is performed which leads to better decision making.

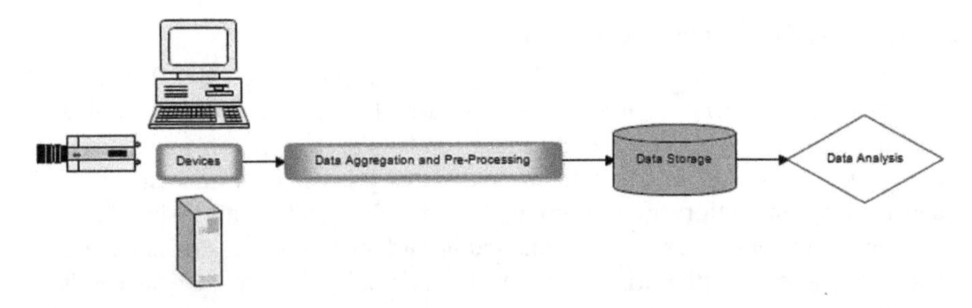

FIGURE 3.1 Demonstration of Data movement.

By using IoT specific products in the healthcare industry it redefines it by the following way:

- Improved Performances
- Better Workflow and Processes
- Ensuring better care
- Reduced cost for Patients
- Improved treatment outcomes
- Proactive treatments
- Error Reduction
- Faster Disease diagnosis
- Drug and Equipment management
- Downsize the amount of waiting time for emergency room
- Tracing patients, staff

3.4 SMART PILLS

Diagnosing a patient with inflammatory intestines unwellness like Crohn's disease (CD) may be a technique that involves multiple steps. If the results of physical examination, blood tests, and stool tests suggest that a person's symptoms are being caused by inflammation within the alimentary canal, then the doctor would possibly refer the patient to a gastroenterologist. Gastroenterologists are physicians specializing within the health of the alimentary canal system; they are specially trained to perform a special sort of the diagnostic procedure mentioned as examination. Examination procedures alter the gastroenterologist to test at intervals a person's alimentary canal using a special instrument mentioned as an "endoscope." associating with instrument may be a long, thin tube with a very little camera and light connected to the top of it, that projects footage of the alimentary canal onto a screen for analysis. Endoscopes are going to be inserted into the body through a natural gap, such as the mouth and down the throat, or through the underside. A medical instrument is also inserted through a little low cut (incision) created at intervals the skin hole surgery is being performed.

The expression "smart pills" refers to smaller than traditional electronic gadgets that are formed and planned within the form of pharmaceutical containers yet perform extremely advanced functions, as an example, detecting, imaging, and medicine conveyance. they will incorporate biosensors or image, hydrogen ion concentration or compound sensors. Once they are swallowed, they move the duct to catch information that is usually exhausting to induce and that they are effectively distributed by the system. Their arrangement as ingestible sensors makes them specific from implantable or wearable sensors. Several endoscopy methodologies and an oversized variety of endoscopy strategies are likewise performed to investigate or screen for cancer. Traditional, rigid endoscope used for these ways are awkward for patients and should cause inward wounding or result in unwellness as a results of utilize on numerous patients. Smart pills wipe out the need for obtrusive systems: wireless communication permits the transmission of continuous data; advanced in batteries and domestically out there memory create them valuable for long-run sensing from within the body

Smart pills have altered the finding of the gastrointestinal issue and will replace standard symptomatic strategies, as an example, endoscopy. Usually, associate endoscope is embedded into a patient's throat, and during this manner the higher and lower duct, for demonstrative functions. There's a danger of aperture or tearing of the muscle system coating, and also the patient faces inconvenience throughout and when the system. A smart pill or wireless capsule endoscopy (WCE), be that because it might, will while not a lot of a stretch be gulped and captive to catch footage, and needs negligible patient readiness, as an example, sedation. The implicit sensors allow the estimation of bodily fluid and gases within the gut, giving the doctor a multidimensional image of the human body.

The expression "smart pills" refers to smaller than traditional electronic gadgets that are formed and planned within the form of pharmaceutical containers yet perform extremely advanced functions, as an example, detecting, imaging, and medicine conveyance. They will incorporate biosensors or image, hydrogen ion concentration or compound sensors. Once they are gulped, they move the duct to catch information that is usually exhausting to induce and that they are effectively distributed with from the system. Their arrangement as ingestible sensors makes them specific from implantable or wearable sensors. Several endoscopy methodologies and an oversized variety of endoscopy strategies are likewise performed to investigate or screen for cancer. ancient, rigid endoscope used for these ways are awkward for patients and should cause inward wounding or result in unwellness as a results of utilize on numerous patients. Smart pills wipe out the need for obtrusive systems: wireless communication permits the transmission of continuous data; advanced in batteries and domestically out there memory creates them valuable for long-run sensing from within the body. Smart pills have altered the finding of the gastrointestinal issue and will replace standard symptomatic strategies, as an example, endoscopy. Usually, associate endoscope is embedded into a patient's throat, and during this manner the higher and lower duct, for demonstrative functions. There's a danger of aperture or tearing of the muscle system coating, and also the patient faces inconvenience

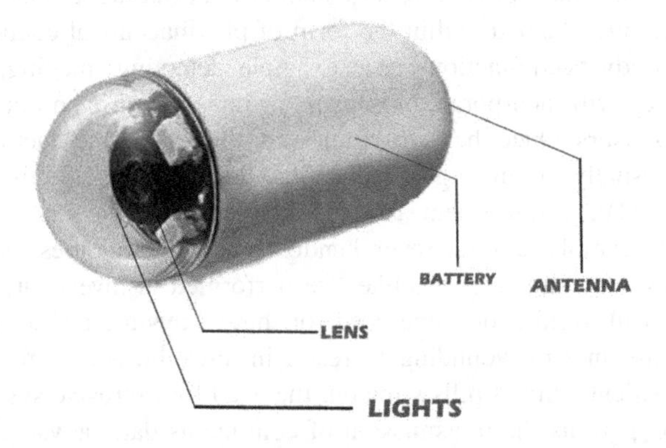

FIGURE 3.2 Internal Structure of Smart Pills.

throughout and when the system. A smart pill or wireless capsule endoscopy (WCE), be that because it might, will while not a lot of a stretch be gulped and captive to catch footage, and needs negligible patient readiness, as an example, sedation. The implicit sensors allow the estimation of bodily fluid and gases within the gut, giving the doctor a multidimensional image of the human body.

3.4.1 Technology of WCE

Wireless Capsule endoscopy or smart pills carries with it three parts

- Capsule
- Information Recorder Belt/ smart wearable
- Workstation

3.4.2 Capsule

The wireless capsule could be a dispensable device, estimating 11 × 26 millimeter (somewhat larger than a considerable nutrient container) and weighing three seven gram and the two vault, chamber form case is formed of a bio-digestible plastic toping with a smooth-surface that permits the body process of the intestinal tract to propel instrumentation through lumen. Wireless capsule contains a complementary metal-oxide silicon chip camera, a lens, illuminating flash-discharging diodes, and vitality supply and radio-telemetry transmitter. Case battery life is around 8 hours, which is adequate for capturing images of the small alimentary canal. At purpose once the battery control is exhausted, the transmitter changes the case into guide mode. While operating in this mode, it transfers data to recorder concerning the world case enabling workstation or framework to check the capsule is doing right for ten further hours. The expendable pills are made of very small bio-digestible material imperviable into abdomen connected liquids. It broadcast video signs and information increasingly. The capsule is eaten by compliant and offers video, photos of gastrointestinal mucous membrane amid it travel at some point of the digestive tract at a rate of two pictures for each second. The system will gain around 50,000 photos. The capsule is generally discharged when roughly 8 to 72 hours.

3.4.3 Information Recorder Belt/Smart Wearable

The wearable wear by the compliant near the midriff gets the sign transmitted by the instrumentation through a spread of sensors set on the complaint's body. The gadget display (antenna) is concerned eight indistinct, 4-centimeter measuring sensors added to the skin by unneeded glue cushions. It gets the pictures from the container related sends a sign to an information recorder. The sensors are identified with the recorder by partner convertible coaxial connection. The data recorded could be a Walkman measure battery-worked unit that gets the information transmitted by the container. It includes a beneficiary, processor modules, and an intense plate drive to store the information. Eight nickel-metal 6-Volt fueled batteries are utilized for the undertaking. The information recorded is prepared for the assignment once the gadget show and batteries and information recorder are out and out related. Blue a flicker light exhibits that the recorder is recording the information. The information recorder will move around fifty thousand photographs to the workstation.

3.4.4 Workstation

Here the system processes the information that was downloaded from the information recorded. The workstation is associate altered standard computer meant for handling of information into motion picture introduction. Here outcomes assures doctors to pursue manner gone the pills, see sores, and spare crucial photo, short-video cuts. Motion picture is enclosed edges it might show one to fifty pictures for each second. It is to prevent and switch skills. a lot of usually than not, the picture is viewed as five to ten pictures for each second except higher or lower velocities may be chosen to work out the images. The doctor examination survey time is around 1.5 hours. Proceeded with advancement and advancement of smart pills has caused the advance in blood detecting calculation that utilizes hemorrhage style, enabling the doctor to concentrate a lot of on hemorrhage areas.[7]

3.5 SMART BEDS

Smart emergency clinics beds have a remote checking framework that monitors the patient. Smart medical clinic beds contain sensors for body temperature, heartbeat, blood, oxygen and weight sensors, among others. Those signs are required and important for specialists to watch the strength of the patients. This strategy is found inside the clinic beds and moves every patient sign to the director, especially in those cases that need restorative consideration. This significant information is conveyed to the focal arrangement of the medical clinic and enables wellbeing managers to in a flash survey and screens the patient's organ. Furthermore, this system sends ready messages, or flag, to the chiefs just in the event of any startling alteration inside the remaining of the patient all through that measure of time. Since a larger than usual amount of your time is spent on the bed, it is basic for the emergency clinics to watch the patient's condition while the individual is on the bed. Smart beds have offered an effective answer for medicinal services providers to unendingly screen patients to create higher consideration to the patients. Such propelled beds without breaking a sweat the weight of the medical clinic specialists that is responsible to watch the patient's condition. Furthermore, smart beds have set up instrumental for social insurance providers to anticipate find and stop unscheduled bed exits.

Emergency clinics beds are any place most patients pay the greater part of their time, all things considered, and new "smart beds," encourage patients to be careful, while "smart" capacities encourage attendants to investigate information and improve patient consideration. The beds interface with EMR systems to send quiet data and encourage medical attendants to screen understanding measurements like development and weight changes. Innovation firms are association inside the smart bed pattern, as well, giving accommodating instruments which will add amicability with the beds to yield even extra significant information.[4]

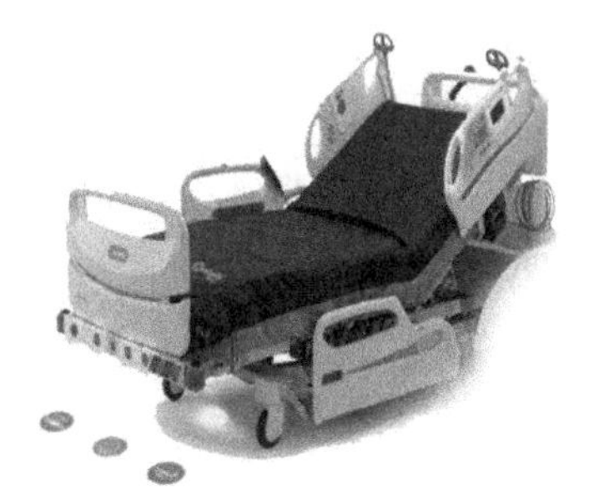
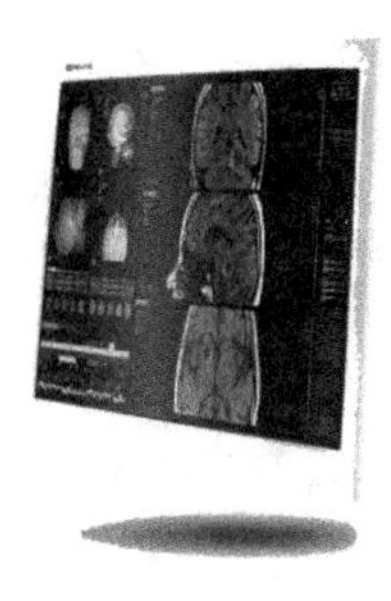

FIGURE 3.3 Demonstration of Smart beds.

One noteworthy focal point of "smart bed" propels are rising patient security and solace all through a likely long emergency clinic keep however quiet wellbeing has perpetually been a spotlight, the sensible Care Act has made patient fulfillment and solace even a great deal of fundamental. With the predominance of patient falls, skin trustworthiness, aspiratory concerns, and hierarchical hazard (parental figure damage) on the expansion, clinics are attempting to discover stock that encourages decrease these dangers and improve results. Falls, explicitly, are a level of concern as opposed to just possessions attendants perceive once a patient is acquiring up, the innovation conjointly makes reports of patient development that medical caretakers will watch for examples.

3.5.1 Features of Smart Bed

A smart bed might be a bed that uses a scope of innovations like weight and heartbeat sensors, the temperature the executives, and auto-lean back to gather data concerning your rest and wake-up examples, that it at that point puts to reasonable use to help you to achieve a greatly improved night's rest.

3.5.1.1 Sleep Tracking

The ability to remain track of anyway well you rest is basically the chief trademark side of an astute sleeping pad as threatening a regular one. utilizing a solitary sheet of "smart texture" or numerous sensors unfurl all through the bed, the smart bedding tracks a spread of information like breath, heartbeat, rest latency and body strain to see an approach to offer you the chief cozy night's rest.

3.5.1.2 Temperature Control

Smart beds for the most part return coordinated with an inbuilt indoor regulator, which grants you to control the temperature of the bedding all through rest.

3.5.1.3 Air Chambers

Expanded air tubes inside the bedding return stuffed with weight sensors that empower it to control your body act though you rest. These air cylinders might be controlled remotely through the smart bed's own application.

3.5.1.4 App Integration

Your smart bed interfaces consistently through the Web of Things to a pile of various smart home devices and applications, together with remote helpers, indoor regulators, keen lights, low makers and TV sets, allowing you to oversee of these totally various gadgets while not regularly getting up.[7]

3.5.2 Technology of Smart Bed

3.5.2.1 Sensor

Exceedingly strong twists sew nylon-polyester mix material that is delicate, flexible, and breathable. It is additionally 100% texture, no bulky or awkward sensors. monitors right around 2,000 weight focuses. Initially intended for hospital beds, the therapeutic evaluation sensor utilizes propelled bedding innovation that guarantees consistent and ongoing weight mapping to lessen weight when required amid the night.

3.5.2.2 Air System

Murmur calm vacuum apparatus. Different air chambers consider unmistakable robustness for different body zones Head, Shoulders, Lumbar, Hips, and Legs.

3.5.2.3 Three Operating Modes

3.5.2.3.1 Programmed

The most creative sleeping cushion innovation work at any point found in a bed. Permits the smart bed to naturally and ceaselessly acclimate to your ideal firmness, whatever your position.

3.5.2.3.2 Position Detection

Set how the firm smart bed feels when you are on your back and how firm the smart bed feels when you are your partner. The smart bed utilizes propelled detecting innovation to recognize your dozing position and will conform to your settings dependent on your development while you rest during the night.

3.5.2.3.3 Manual

Adjust support for each zone of your body: Head, Shoulders, Lumbar, Hips, and Legs. Empowers you to change the vibe as normally as your needs change.

3.5.2.4 Network

Announcing abilities to show rest information on tablet (included) or different remote gadgets.

Wi-Fi availability that permits smart bed to coordinate with existing and future advancements.

3.6 SMART WEARABLE

Wearable innovation could be a sign of the Web of Things and furthermore the most inescapable of its usage up until this point. The strength of data procedure accomplished by various smart wrist wear, savvy pieces of clothing, and restorative wearable is going to the reason any place this shopper situated aspect of the IoT innovation can carry the excellent cost to our lives and become a fresh out of the plastic new style on the methods. The IoT Platform improves wearable innovation with excellent prepared to-utilize IoT capacities and applications. IoT is certainly incorporated with little microchips in wearable gadgets and grants moment capacity, a gadget the board, data combination, occasions and standards execution, security, and various choices. IoT conjointly gives versatile cloud common sense to affirm misfortune free correspondence between wearable gadgets and to engage them with learning examination and visual picture apparatuses. Fundamentally, it is not exclusively the client who will like propelled data investigation anyway conjointly gadget creators, who may follow execution and gadget wellbeing details, dissect the precision of finder calculations, and execute A/B testing for the different gadget programming framework. Wearables' innovation could be an incredibly adaptable IoT stage-structured on contemporary smaller scale administration plan that offers a boundless opportunity of customization, an option of advancement devices and dialects, and a kind of cloud preparing. Subsequently, your group will rapidly begin with wearables and abbreviate the occasion course of events for cutting-edge wearable applications to weeks or perhaps days. Planned with IoT, wearable applications emerge from the past age of simple arrangements and convey everything that a well-informed customer may anticipate.[2,5]

3.6.1 Smart Watch

Wearable advancement is a keen electronic device with microcontrollers. The device can be worn on the body as an inserted or as an extra. These wearable contraptions can play out an extensive parcel of indistinct enrolling assignments from cell phones and PCs, now and again, wearable advancement can beat these hand-held devices totally. While wearable advancement will when all is said in done, insinuate things which can be put on and taken off effectively, there are progressively prominent adjustments of the thought as by virtue of installed contraptions, for instance, microchips or even sharp tattoos. In the end, paying little heed to whether a device is worn on or joined into the body, the purpose behind wearable advancement is to make steady, worthwhile, reliable, helpful, and generally without hands access to equipment and PCs.

Smart watches are computerized watches that achieve more—significantly more—than your old straightforward time GPS guide. We are not talking about that once-dumbfounding calculator watch that you had in evaluation school. These are verifiably best in class gadgets. Smart watches can run applications and playback a wide scope of computerized media, like soundtracks or radio spouted to Bluetooth headphones. Countless these watches have touch screens, which empower you to get to limits like a calculator, thermometer, compass, and anything is possible from that point. Most of the present cycles of smart watches are not completely free contraptions, just in light of the way that they miss the mark on an internet affiliation. Such an enormous number of the watches are proposed to interface directly with various devices that do have internet accessibility, to be explicit to your cell phone.

Essentially, as with your cell phone, the internet gets to engages a smart with the whole universe of potential limits, as message sees, GPS course, and logbook synchronization. In addition, clearly, a Bluetooth relationship with your phone suggests the watch can empower you to put calls or send and get messages. Some smart watches are made expressly for recreations purposes, allowing you to pursue your lap times, partition and course. They may work pair with additional items, for instance, a heartbeat screen or rhythm sensor. There are the distinguishing strength smart watches manufactured especially for cruising aficionados, helping them track factors, for instance, speed, wind heading, and wind speed.

A smart watch is a wearable innovation as a wristwatch; present-day smart watches give a close-by contact screen interface to step by step use, while a related cell phone application obliges the officials and telemetry. While early models could perform principal assignments, for instance, calculations, propelled time telling, elucidations, and delight playing, 2010s smart watches have progressively expansive convenience closer to cell phones, including compact applications, adaptable working frameworks, and Wi-Fi or Bluetooth organize. Some smart watches function as helpful media players, with FM radio and playback of cutting-edge sound and video records through a Bluetooth headset. A couple of models, called 'watch phones' have adaptable cell convenience like making calls. The product may consolidate propelled maps, calendars, and individual facilitators, number crunchers, and various sorts of watch faces. The watch may talk with external contraptions, for instance, sensors, remote headsets, or a heads-up introduction. Like various PCs, a smart watch may accumulate information from the inside or outside sensors and it may control, or recuperate data from, various instruments or PCs. It may support remote advancements, for instance, Bluetooth, Wi-Fi, and GPS. For certain reasons, a watch computer fills in as a front end for remote frameworks, for instance, a cell

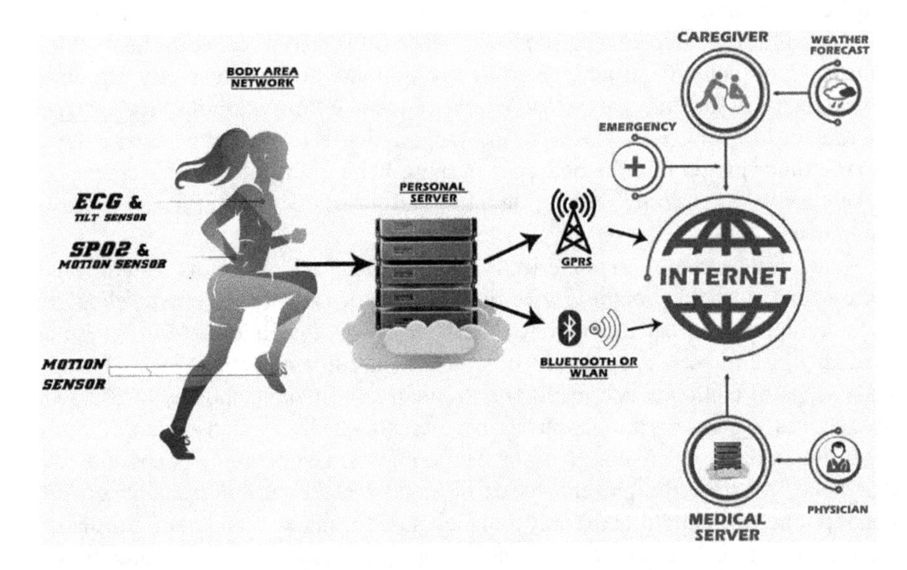

FIGURE 3.4 Influences of IoT and Wireless sensors in day-to-day life.

phone, talking with the PDA using distinctive remote advancements. Smart watches are advancing, especially their arrangement, battery point of confinement, and prosperity related applications. [2,5]

Smart watches are responsible for the following things:

- Keep track of day-by-day practice and calorie intake.
- Get wellbeing cautions identified with glucose levels and other indispensable signs, especially those identified with particular conditions with which the wearer is afflicted.
- Catch biometric information, for example, pulse, oxygen levels, blood pressure and other important wellbeing information.
- Send gathered information to medicinal services suppliers.

Access investigation and announcing that gives input dependent on perceptions of information gathered after some time.

3.6.2 How Wearable Works

Wearable depends on three layers:

First Layer – This layer consists of sensors. The sensors are set nearest to the body. The sensors monitor components, for example, temperature, motion and heartbeat.

Second Layer – This layer is responsible for the availability and control layer. The Bluetooth Low Energy convention is the most regularly used to associate wearable gadgets to a Smartphone or home system.

Third Layer – This layer is the place where the wearable supplies and peruses information.

There are several types of smart watches available now. Some of them are listed below.

- Apple Watches
- Tizen Watches
- Wear OS Watches

3.7 REMOTE HEALTH MONITORING

Observance programs will gather a decent fluctuate of wellbeing learning from the reason for consideration, as significant signs, weight, weight-level, blood glucose, blood component levels, pulse, and electro-cardiograms. This information is then transmitted to wellbeing experts in offices like watching focuses on therapeutic guide settings, clinics, and medicinal guide units, adaptable nursing offices, and concentrated off-site case the board programs. Wellbeing experts screen these patients remotely and follow up on the information got as a piece of the treatment set up. Observing projects may likewise encourage keep people sound, empower more seasoned and debilitated individuals to quantify gathering longer and abstain from moving into flexible nursing offices. The rate may likewise serve to reduce the number of hospitalizations, readmissions, and lengths of keep in medical clinics—the majority of that encourage improve personal satisfaction and contain costs.

Notwithstanding raising the quantity of consideration, the rate also has the adaptability to support the standard of consideration. Since RHM associates clinicians extra legitimately (and about right away) with applicable patient data, it makes their day-by-day schedules extra conservative and facilitates the probability of burnout prompting clear points of interest to patient consideration. Considerably higher, RHM improves understanding conduct by making a framework by which people are also drawn in with, and in order of, their wellbeing. Viable RHM projects supply innovation that far from being the chilly, clinical, overwhelming therapeutic innovation that springs to mind for a few customers is extremely cozy and familiar to patients. Significant as patient solace and commitment seem to be, the benefits of remote patient viewing rise above that, giving patients the profitable confirmation that someone is taking a look at his or her wellbeing and prosperity on multi-day today.[3]

3.7.1 Technological Components

Remote Patient Monitoring systems consists of the following parts.

- Sensors embedded in a gadget which is empowered by remote interchanges for measuring physiological parameters.
- Information which is kept locally, information at compliant site interfaces among sensors other centralized-information store also additionally human services suppliers.
- Brought together vault to store information convey sensors, resident information storage, diagnostic-applications, as well as health care insurance suppliers.
- Diagnostic-application programming that creates treatment proposals and intercession alarms dependent on the examination of gathered information.

Some examples of RPM technology include:

- Glucose meters for patients with polygenic disorder.
- Heart rate or pressure level monitors.
- Continuous surveillance monitors which will find patients with conditions like dementedness and alert health care professionals of an occurrence sort of a fall.
- Remote sterility treatment and observation.
- At-home tests which will keep misuse patients answerable for and on target with their goals.
- Caloric intake or diet work programs.

3.7.2 How RPM Works

While RPM strategies can change contingent upon the gadget being utilized or the condition being checked, a large portion of the innovation incorporates comparative parts. The main is a remote empowered sensor that can gauge explicit physiological parameters and store the information it gathers. This storage must likewise incorporate an approach to interface with extra sensors, human services supplier databases and related applications. Applications commonly give clients an interface to follow or break down the information and show treatment proposals.

The information gathered by RPM gadgets is sent to the best possible area and put away in a social database. This permits human services associations with remote broadcast communications information to be taken a look at as individual occurrences or with regards to a whole health history. Regularly, the gadget can alarm patients when a human services supplier has investigated the information or identifies an issue that requires the patient to come in.[3]

3.7.3 Benefits of Remote Patient Monitoring

- Quickly access to medicinal services
- Improved nature of consideration
- Peace of mind and everyday confirmation
- Improved backing, instruction, and input
- Increased patient engagement

Some of the Top RPM companies are listed below.

- Dexcom
- Honeywell Life Sciences
- Medtronic
- Philips Healthcare
- ResMed
- Senseonics

3.8 IOT-ENABLED APPLICATIONS

Remote monitoring is very helpful for the senior, immobile, inveterately sick, or folks living alone. The technology uses IoT-enabled daily observation devices like pressure monitors, heart monitors, or glucose meters to gather relevant information and build it accessible to the physician. He/she can, thus, closely monitor the health of the patient and intervene once needed.[4,5]

- Philips e-Alert is associate IoT-enabled tool that monitors vital medical hardware like MRI systems and warns health care organizations of an imminent failure, preventing uncalled-for period.
- IoT devices labeled with sensors are utilized in hospitals to stay track of medical instrumentation like nebulizers, wheelchairs, and defibrillators. The real-time locations of medical employees are equally monitored and analyzed.
- IoT devices are valuable in hospitals for infection management, pharmacy inventory management, and environmental observation like temperature and humidity.

- Health insurance corporations could realize information capture by IoT-enabled wearable helpful for detection frauds and corroboratory claims.
- Wearables like Fit Bit that tracks many health parameters, AmpStrip that monitors pulse rate, smart watches that find apnea, and good glasses to assist the blind are greatly dynamic the method the health care trade functions.
- QardioCore, an graphical record monitor designed to produce continuous medical-grade knowledge. Users will wear this device inside their normal lifestyle: at work, the gym, or out and concerning. The information is alleged to assist patients to raised monitor health conditions like high blood pressure and cholesterol. It additionally sends data into health centers that monitor conditions like polygenic disease, heart troubles, and weight gain, while not the necessity for physical visits.
- Zanthion may be a medical alert system that is worn by a patient as an item of consumer goods or jewelry. It feeds into a series of connected sensors that live the health and welfare of the user. If a patient were to fall out of bed or stay nonmoving for too long an amount, an alert is distributed to family or friends who will facilitate.
- Screen Cloud is being employed by hospitals and tending professionals already, with applications that are thinking so much outside the box of traditional digital accumulation. Take the organization mistreatment video art and digital accumulation to boost patient welfare in hospitals, with established effects of lower stress levels and anxiety in patient waiting rooms.
- Up by Jawbone may be a fitness tracker with a distinction. Instead of simply count calories and steps, it will be accustomed monitor all aspects of health, from weight and sleep patterns to activity and diet, to permitting the user to create higher health selections supported knowledge. Some patient reform teams are even mistreatment it as some way to support those with weight and health problems from afar or outside of the physical heart itself.
- Wireless sensors like those from detector Matrix are being employed in hospital refrigerators, freezers and laboratories to confirm that blood samples, medications, and alternative materials are unbroken at the right temperature.
- NHS take a look at beds are smart, connected beds being employed within the UK's NHS system that monitor patients and track knowledge. Combining wearable monitors with alternative knowledge detector sources, they save cash and time, permitting aged patients and people with long-run health conditions to observe their progress and problems a lot of expeditiously.
- Swallowable sensing elements are just about what they say on the box: some way for patients to avoid colonoscopies by swallowing a sensor the dimensions of a cod liver oil pills. This sensing element will diagnose issues encompassing conditions like irritable viscous syndrome and carcinoma in place of a lot of invasive surgeries.
- Propeller's Breezhaler device may be a connected sensor that creates the management of bronchial asthma or COPD easier. The sensor attaches to the highest of the pump and monitors knowledge whenever it is used. The user to gather information on triggers and enable relations and doctors to manage use,

all from a mobile app. It is the same to cut back the amount of bronchial asthma attacks and increase the number of symptom-free days.

3.9 CONCLUSION

We have discussed the various types of application of Internet of Things (IoT) in the healthcare industry in this chapter. Then we discussed the smart bed technology, Wireless capsule endoscopy (WCE) or smart pills, smart wearable like smart watch, Remote health monitoring, IoT-enabled applications and how this technology works and how it is beneficiary for the patients in the industry of health care. As discussed above all the benefits associated with the above all applications it can be seen that using IoT in the health care industry overcomes the problems related to it. Thereafter the concept of bright future of IoT and its contribution in the healthcare industry were briefly discussed.

REFERENCES

1. Alok Kulkarni, Sampada Sathe "Healthcare applications of the Internet of Things: A Review" *International Journal of Computer Science and Information Technologies*, Vol. 5 (5), 2014, 6229–6232.
2. B. Sobhan Babu, K. Srikanth, T. Ramanjaneyulu, I. Lakshmi Narayana "IoT for Healthcare" *International Journal of Science and Research (IJSR)* 2014.
3. Shubham Banka, Isha Madan and S.S. Saranya "Smart Healthcare Monitoring using IoT" International Journal of Applied Engineering Research ISSN 0973-4562 Volume 13, Number 15 (2018) pp. 11984–11989.
4. Ravi Kishore Kodali, Govinda Swamy and Boppana Lakshmi "An Implementation of IoT for Healthcare" 2015 IEEE Recent Advances in Intelligent Computational Systems (RAICS) 10–12 December 2015.
5. Stephanie Baker, Wei Xiang, Ian Atkinson "Internet of Things for Smart Healthcare: Technologies, Challenges, and Opportunities" DOI 10.1109/ACCESS.2017.2775180, IEEE
6. www.peerbits.com/blog/internet-of-things-healthcare-applications-benefits-and challenges.html
7. www.wipro.com/en-IN/business-process/what-can-iot-do-for-healthcare-/
8. www.i-scoop.eu/internet-of-things-guide/internet-things-healthcare/
9. https://internetofthingsagenda.techtarget.com/feature/Can-we-expect-the-Internet-of-Things-in-healthcare
10. https://searchhealthit.techtarget.com/essentialguide/A-guide-to-healthcare-IoT-possibilities-and-obstacles
11. https://econsultancy.com/internet-of-things-healthcare/
12. www.iotforall.com/exciting-iot-use-cases-in-healthcare/
13. https://bmcmedinformdecismak.biomedcentral.com/articles/10.1186/s12911-018-0643-5
14. www.dotmed.com/news/story/22656/
15. www.lifewire.com/smart-bed-4161313
16. www.mouser.in/applications/article-iot-wearable-devices/
17. www.telit.com/industries-solutions/healthcare/wearables/

18. www.inxee.com/Smart-Wearables.html
19. www.hindawi.com/journals/js/2018/6272793/
20. https://internet-of-things-innovation.com/insights/the-blog/iot-trend-watch-wearable-technology/#.XLnydL7hXIU
21. https://en.wikipedia.org/wiki/Smartwatch
22. www.ronpub.com/OJIOT_2018v4i1n07_Ngu.pdf

4 Putting Blockchain into Practice

Preet Deep Singh
Invest India

CONTENTS

4.1 WHAT ARE WE TRYING TO SOLVE?

Blockchain is a distributed immutable ledger that is encrypted. It is immutable for two reasons: time stamp and the encryption of preceding content. This looks simple enough. On face value, this does not solve much. Google sheets can be tweaked to become an immutable ledger with open access. We only need to find a way to introduce one-way encryption so that only the owner of the original node knows the verification.

It is surprising how hard it is to find all these in one system. Decentralization has been made possible in the past decade due to better internet penetration. Earlier, it would not be accessed in real time. Encryption has become very important in the past few decades and privacy is being equated with a fundamental right. People would not want someone to know their health records, their expenses, their purchases and so on. Some western jurisdictions have taken a very strong stance toward privacy and penalized corporates for their lapse.

Cambridge Analytica was a famous scandal highlighted in the light of the US Elections in 2016 where data from Facebook and other sources was used to target advertising in a way that it is said to have affected the outcome of the elections.

Opportunity flows from needs. Privacy is a need and cryptography is a solution. To recall, cryptography is one-way encryption. This means that all areas where privacy is required are places where blockchain has a natural application. Similarly, any place where ledgers are involved would have an application of blockchain. These cases might or might not require encryption. Immutability provides trust to any transaction. A buyer does not have to trust the seller (and with encryption, the buyer does not even need to know the seller). Transparency is needed wherever there is money involved and when the rest of the process is not in the hands of the initiator.

From this we deduce that health, finance, real-estate and governance can unlock value through implementation of blockchain.

Blockchain is at the back end in any process and no one gets (or needs) to see it. To understand this better, turn to WhatsApp. When you send a message to a new person you get a message saying that all messages are encrypted. This does not mean that the person at the other end cannot understand your message. It means that during transmission it is encrypted. Communication between the receiver and sender is not different, it is only the journey through the medium that is encrypted. Similarly, in the case of blockchain, the components and user interface are not affected by this. Just the log of all transactions is maintained on the blockchain and one can be assured of privacy.

4.2 SOME USEFUL DISTINCTIONS

4.2.1 Difference between Blockchain and Cryptocurrency

One is to put things in an immutable way, the other is to have that as currency. Cryptocurrency, as we know it, is currency that is based on the blockchain.

These can be understood as concentric circles where cryptocurrency is a token based on blockchain.

4.2.2 Public Versus Private

A public blockchain is one in which anyone can participate in any capacity whereas in a private blockchain the owner of the blockchain defines who all can participate and in which capacity. This means private blockchains are 'permissioned' blockchains where only participants selected by the owner of the blockchain can verify the information.

Since public blockchains involve anonymity, most of the blockchain solutions that are being developed for internal use by corporates would be private blockchains. Solutions for governance would have a combination of both where some data is available to the public and some other data is accessible based on permissions.

4.3 POSSIBLE APPLICATIONS

4.3.1 Finance

Since blockchain shot to fame with the bull run of bitcoin, the first application of blockchain is that of payments.

4.3.1.1 Know Your Customer

According to the Report of Steering Committee on Fintech[1] putting Know Your Customer (KYC) and Anti-Money Laundering (AML) systems on blockchain would reduce costs and effort by banks to comply with these requirements.

Similarly bill discounting could be made a lot easier. If a manufacturer has a work order, they could put it on the platform and a bank could verify the work order and discount the bill, providing finance to the manufacturer. In case of default the record of the buyer would reflect this and would be for all to verify. These records would also allow people to have a credit history, which is valuable for lending institutes.

4.3.1.2 Cross-Border Transfer

If you want to transfer money to a relative in another country, the payment is routed through multiple banks and intermediaries that receive, collate, net-off and transfer the money. This leads to a long time and consequently high cost in such a transfer. Application of blockchain in these cases would lead to intra-day and inter-bank liquidity. Use of digital tokens would further expedite the payment process but this assumes cryptocurrency (and not just blockchain). Some banks have worked with Ripple to explore inter-bank-cross-border transactions. Ripple is a cryptocurrency that reduces settlement times to less than a few minutes from a couple of days.

4.3.1.3 Securities Transactions

Stock market trading is famous for T+2 settlement and T+0 settlement. T+2 means it takes securities two days after the day of trade to reach the account of the buyer. Similar T+0 means the securities are transferred to the account of the buyer on the same date. T+0 is available mostly in cases where the buyer has a bank account with the intermediary. This way the monetary transaction and the reverse securities transaction are done by the same intermediary for the same person. This cuts down on the KYC time. Application of blockchain would radically decrease settlement time for securities.

4.3.1.4 International Trade

Payments in international trade are the most important because of lack of trust between traders. Banks through 'Letters of Credit' (LC) bridge this gap to an extent. The Punjab National Bank Scam[2] involving Nirav Modi highlighted the lack of reliability of these LCs. Blockchain is a secure and fast way of ensuring that the issuing bank and the honoring bank would know about the intent and actions of relevant members.

4.3.2 Smart Contracts

A smart contract is a way by which conditions are fed to a machine and money is blocked therein. When the coded event occurs, payment is automatically transferred. This is similar to what PayPal was for e-commerce. Since all this is on a blockchain, there is little doubt about the payment. An external and independent source confirms the occurrence of the event. This source is called Oracle. A service that tracks flights is an Oracle in case of flight insurance. Similarly, the Met[3] declares certain districts as flood affected or drought affected. In case of crop insurance this could be used to trigger compensatory payments by the government.

In case of sports betting, the payment risk is assumed by the broker. If someone were to use ESPN or another sports application as an Oracle, then all wagers could be coded as smart contracts and counter-party risk would be assayed.

The scope for blockchain implementation in this is huge. Contract-related litigation is a huge component of Ease of Doing Business in any jurisdiction. Contracts have two sides, performance and payment. If either or both of those can be secured then the other can be tamed. In case of smart contracts, the payment is blocked and the performance is evidenced by an Oracle. This would lead to fewer disputes causing more and better trade.

Litigation, non-execution, non-payment. If we guarantee one side then the other would be solved automatically.

4.3.3 Property

Property Records are the most precious possession of most citizens. Any error in these could lead to revocation of property rights, problems in insurance claims and at time of sale/hypothecation. There is only one ownership record for all properties, and that is with the government. The process of change in ownership of property through sale/purchase/mortgage and the process of change in records are each carried out in different offices and come under different departments in certain cases. While property sale/purchase is a revenue item, that of land ownership is a record item. This divorce further adds to incorrect and incomplete updating of records.

Any error or malfeasance on part of someone with access to ownership records can lead to a lot of monetary loss and frustration. It is not easily possible to trace these changes to correct these. Title suits can take years to get resolved and in the interim period the property is useless.

If the two processes could be linked in a secure and verifiable way, it would solve a lot of problems. While some states have tried to link land registry through digitization and making the records open, blockchain can add a layer of security for land owners.

4.3.4 Healthcare

Having and maintaining medical records is beneficial for patients. They help doctors diagnose things better and prescribe a smoother medical route. It would lead to fewer experiments (by knowing what the patient has already gone through). Tracking of blood pressure and heart rate in our phones can provide a snapshot of the patient's wellbeing to a trained doctor. Family history would allow a doctor to target tests better and to monitor progress.

Accumulation of these benefits is prevented by privacy concerns associated with Electronic Medical Records. Health data is used by companies for designing solutions and to study the impact of interventions vis-à-vis a control group.

Data can be used for marketing new drugs to people which is at the boundary of ethically accepted practices. This data can also be used to malign someone or to prevent them from taking up public office. Because of all these use cases and concerns regarding medical data its cost on the black market is very high. Anecdotal evidence[4] suggests that medical data is thousands of times costlier than credit card data. This reflects a clear gap in the market. If there were a secure way for people to share anonymized data in exchange for money or credit (that can be used on the platform, in the form of tokens) it could be a win-win situation.

De-identification of data is a key component. Blockchain would ensure anonymity and control. The patient can decide who can view what component of the report and whether the updated content would be accessible. Cryptocurrency can ensure that people who consent get paid for sharing it.

4.3.5 Insurance

Insurance is one industry that requires a lot of documentation and verification. This is done multiple times. Insurance companies go to great lengths to ensure that they are paying the right person with the right set of documents for a covered event that happened during the insured period. All this verification takes time and effort. Efforts duplicated at different levels to ensure there is no wrong-doing on the part of the insured, inspector, processing staff and others.

Car insurance typically involves the following steps:

- Insurance cover
- Event which in this case is an accident
- Claim
- Claim verification

- Claim report
- Damage report
- Internal processing
- Bill submission by mechanic
- Verification
- Payment

For each claim, the company wants to assess

- Whether the same car was insured
- Whether the driver had a valid driver's license
- Whether the documents of the car are in order
- Whether a police report, if needed, had been filed
- Whether this damage existed before the insurance period
- Whether this damage existed after the insurance period
- Whether an independent party has seen the claim[5]
- Whether the correct claim has been filed by the mechanic
- Whether the part that had to be repaired was replaced or repaired
- Whether the prices of the parts are as per notified prices
- Whether the bills are for the right car
- Whether the dates on the bills and those on the claim match

Assessing all these basic things requires matching and verifying documents. This is subject to human error and malicious intent in some cases, to counter which it happens at multiple levels. While these checks seem simple, they are not so. Also, the number of claims processed by an insurance company is too huge for it to be done efficiently. There are anecdotes that a public sector general insurance company took over three days to verify that the vehicle for which a claim was made was in fact covered by the company.

In case this data were available in a digitized format, verification could be coded and conducted effortlessly. On blockchain, this could be immutable and distributed, which would ensure transparency.

4.3.6 Supply Chain

When goods are moved from the producer to the user, or when high-value items are resold, one major concern is their genuineness. Organic fruits and vegetables command a premium. The resale value of a Rolex is very high. Wine from certain vineyards is valued more. How can the end-user be assured of the origin of these products? Blockchain is being considered as a major solution.

If at each stage of processing or change in ownership through agents and transporters, the data is fed on to a blockchain, the end-user can simply scan the product and verify the origins. This would require the producer to enter all relevant details of the product on the blockchain and affix a code or engrave a microcode on the product that can be scanned and the information would be made public. This makes blockchain a good application wherever counterfeit goods are a problem. This extends to pharmaceuticals as well as durables. Companies such as Walmart and Pfizer have conducted successful pilots with this technology.[6]

4.3.7 Governance

Government is the biggest spending entity in the country. Most of it is mired in processes. Tracking that money is a headache for most government departments themselves. The subsidies and grants disbursed by the Central Government pass through multiple hands before reaching the beneficiary. Blockchain can enable tracking of government funds and it is open to audit at all times by anyone who may wish to see it. Any claims made by beneficiaries can be quickly resolved by verifying them on the blockchain. Transparency and immutability would be very useful in ensuring that the money reaches the right people.

4.4 GOVERNMENT'S TAKE

The government is yet to formulate a concrete policy on the issue. There have been two reports that have examined blockchain and cryptocurrency. There is consensus on the use of blockchain and exploring its application in governance. In the 2018 Budget speech, the then Finance Minister Shri Arun Jaitely clarified the stand of the government in Parliament.[7] He said a distributed ledger system or the blockchain technology allows organization of any chain of records or transactions, without the need of intermediaries. The government will explore use of blockchain technology proactively for ushering in a digital economy.

The contentious point was regarding cryptocurrency. The Government constituted an Inter-Ministerial Committee (IMC)[8] in November 2017 under the Chairmanship of Secretary, Economic Affairs comprising Secretary, Ministry of Electronics and Information Technology, chairman of Securities Exchange Board of India and other members to study the issues related to virtual currencies and propose specific action to be taken in this matter.

4.4.1 Summary of IMC Report

While some countries have outrightly banned cryptocurrencies and some others have used them for barter transactions and even as a means of payment, no country has yet considered cryptocurrencies as a legal tender.

The Inter-Ministerial Committee (IMC) set up by the Indian Government to study the issues related to virtual currencies and propose specific actions,
recently recommended that all private crypto currencies should be banned in India except the ones issued by the Government.

While the Committee suggests that the Government keep an open mind on digital official currency and has highlighted the positive aspects of Distributed Ledger Technology (DLT) and its uses in banks and financial firms, it has proposed banning of all private cryptocurrencies given the risks associated with them.

The disadvantages of virtual cryptocurrencies have been further explained below with sources and rationale.

- Public blockchain needs decentralized verification such as in the case of bitcoin. People who verify the transaction and get some bitcoins are called miners. A complex mathematical problem is posed to verifiers and solving these problems takes up a lot of energy. Decentralization may help in keeping this process less cumbersome.
- Low scalability: To be used as a means of payment, the number of transactions that can be processed has to be much higher. At its peak, it would take bitcoin more than a day to verify a transaction.
- High cost: The cost of verifying transactions through the Proof of Work protocol is very high. The report says that 19 US households could be powered for a day by the electricity it takes to mine one bitcoin. If such currency mining is not prohibited in India, it would be catastrophic. Developed nations such as Canada have had to buy power in the open market. India is already power starved.
- Irreversibility: All transactions on blockchain are irreversible. While this gives it credibility, it penalizes mistakes. Anyone who has worked at a bank would tell you that mistakes are common even while transferring money.
- Security: Wallets and even exchanges have been prone to cyber-attacks thereby causing a security concern.
- Password: If one forgets the Private Key, there is no recourse. Currently, we have the provision of verifying the identity of the person and regenerating a PIN or an OTP. No central authority exists in case of cryptocurrencies.
- Monetary policy: cannot be enforced by central banks. This can cause unchecked inflation.
- Cross-border control over currency: cannot be exercised as there is no central clearing house.
- Anonymity: in transactions is causing cryptocurrencies to be used for criminal activities; from financing narcotics, as in the case of Silk Road, to terrorism.

Given the disadvantages, IMC proposes the enactment of a law to prohibit trading and mining cryptocurrencies and a fine of up to Rs 25 Cr and imprisonment of as much as ten years for anyone dealing in them. However, it does recommend official digital currency with the status of a legal tender and appropriately regulated by the Reserve Bank of India.

4.4.2 Report of the Steering Committee on FinTech Related Issues

The report of the Steering Committee on FinTech Related Issues, also within the Department of Economic Affairs, Ministry of Finance has taken a much softer view on the issue. It has encouraged the use of blockchain and has noted global interest in cryptocurrency. They acknowledge that virtual currencies through Initial Coin Offerings are revolutionizing fintech. it also talks about utility tokens and their potential to enable various industries.

4.5 SUCCESS STORIES

4.5.1 LaVis Wine

Counterfeit is a major problem for the wine industry. People want to be assured of what they are drinking especially when they are paying so much for each bottle. Ernst and Young (EY) has been the blockchain implementation partner for LaVis. LaVis is claimed to be one of the first to sell blockchain-certificate wine using EY Ops Chain[9]. The blockchain contains all details related to the story of the wine starting from the type and quality of the grapes, to the date of bottling and the quality of the sulfates. Each time the bottle changes hands between the producer to the wholesaler to the retailer, the status is updated.

4.5.2 TReDS Bill Discounting

TReDS or Trade Receivables Discounting System which is an RBI[10] approved platform for discounting bills for MSMEs, has started using blockchain[11]. This allows them to broadcast information across a shared network where not only is all information in the public domain, it is also sensitive as it deals with financials.

Blockchain allows them to operate while protecting privacy.

4.5.3 Travel Insurance

Axa has partnered with Fizzy to provide travel insurance on blockchain. As mentioned above, blockchain is at the back end and the insured have little or no visibility on that. The process of getting flight insurance or to get any insurance for that matter has always been easy. The problem is at the time of claims. Blockchain is solving that component. Insurance pays off in the event of that condition being met. The insured is supposed to file a claim and then that claim is processed internally by the insurer and payment is released. In case of flight insurance, the insured and the insurer check as to whether the flight was delayed by two hours or more, or cancelled altogether. The advantage in this case is that as soon as a flight is delayed by two hours, the information is relayed at the same time to the insured and the insurer. And since it is coded on Ethereum smart contracts, that information triggers payment of the sum insured without any human intervention. Even if the insurer wants, they cannot edit the contract once it is on the system. The contract address is: 0xdc3d8fc2c41781b0259175 bdc19516f7da11cba7 which can be accessed on etherscan.io and other such options for Ethereum Blockchain. Anyone can view the contract and the code.

The link to the website is https://fizzy.axa/en-gb/

Other good insurance use cases can be found at https://builtin.com/blockchain/blockchaininsurance-companies

4.5.4 Sweden Land Records

Since land record titles are susceptible to minor forgeries and clerical errors, the Swedish government has decided to experiment[12] with the use of blockchain. It is called Lantmäteriet. Instead of signing papers, people could use digital signatures. Since these can be validated online, a transaction can be concluded. The system operates on a private blockchain and includes land registry and other parties that hold land records such as banks. However, even this system does not act like cryptocurrency where high-value transactions are easy. Nevertheless, just by removing paperwork and preventing forgery, Sweden would save more than $100 million.

4.6 CHALLENGES

4.6.1 Digitization

Driving licenses can be put online in a cryptic form only to check validity and if there have been any challans in the past due to traffic violation. This would allow insurers to decide premium based on the driving history of the person. The problem with this solution is that all the driver's license data is not available in one online database that is accessible and compatible with other databases.

Firstly all data would have to be digitized with a collaborative architecture. This is an exercise that has been undertaken in many developed economies and efforts for this are under way in India as well.

4.6.2 Property Records Complete

Property records on blockchain would ensure that there is no amendment to land ownership without consent (and in some cases, corresponding payment to the seller). The problem with this approach is that it is only prospective. In order to be retrospective, the land records have to be clean in the first place. In the absence of this, the legacy data cannot be digitized on to blockchain. While this may seem very simple it is not so. Even some of the best States in Government of India's Ease of Doing Business Ranking do not have even two years' worth of property records and transactions digitized[13]. Property is fraught with disputes, some in civil courts and others in criminal courts. It is mortgaged, pledged and rented. There are cases subject to Wills, Gift Deeds and Sale Deeds. This leads to too many ownership disputes. If this is not free of errors and disputes, it cannot be put on the blockchain as they would then be immutable.

4.6.3 Need for Scale

The benefits will only arise once the entire platform is on the system. We need everyone to use it. Imagine using emails only to find out that you need to take a print of each communication, respond in writing and each response will then be digitized. It would not add any value, right? Similarly blockchain will add value if all processes are present on it. Therefore the need for scale is high to unlock value.

4.6.4 Coders

The main platforms at the moment are

1. Hyperledger: by IBM
2. Corda R3: industry collaboration initiative with leading organizations as partners
3. Solidity: Ethereum blockchain
4. Others such as MultiChain, SmartChain and other chains: platforms to create your own blockchain

While Hyperledger and Corda require mostly Java knowledge, Solidity and other platforms require more sophisticated knowledge. There is a dearth of people who understand blockchain architecture and can implement it. Coders are in short supply for the entire industry but more so for blockchain.

4.6.5 Smart Contracts

Oracles: smart contracts require Oracles. These Oracles are assumed to be sources of truth for a contract to get executed. For smart contracts to be a reality, we would need many more Oracles. We would also need more reliable and consistent Oracles. For example, flight.com is a reliable source for flight data. However, it can get hacked. There might be an internal maintenance issue. The server might be down or it could have bugs. Since it is owned privately, one cannot rule out any interference. These issues hamper trust.

Non-repudiability of smart contracts is a welcome feature in many cases. However, in certain cases it can prove to be problematic. If someone were inebriated or under duress while they were writing a smart contract, it could have difficult consequences. The best software from the biggest companies also suffers from bugs. It is inconceivable that no smart contracts would have bugs. Further, both parties may agree to dissolve the contract, or there could be a *force majeure*. In these cases we would need an escape chute from the non-repudiable smart contract which is not possible currently.

4.7 DO IT YOURSELF

4.7.1 HyperLedger

Click and Build platforms such as Hyperledger by IBM are available free of cost for enthusiasts. The online program is free for all and leads to a decent working knowledge of the platform. The course helps you develop a small blockchain for trading cars. It does not require any coding experience. In order to build something useful for an organization, it has a Java component that needs to be understood and coded. However, this is a good place to start if you want to understand and see it in action.

Coursera and a number of other learning platforms provide free learning content hosted at www.coursera.org/lecture/blockchain-platforms/hyperledgerpart-1-87Qz4.

4.7.2 Bitcoin

Although Bitcoin and trading in any form of crypto is not recommended in India, it is possible to look at the blocks. Since it is open, you can look at all the blocks ever created. The hash of each block is available. If you click on one of these, you will see the number of transactions in each block which contains the amount of bitcoins transferred, the sender's address and the receiver's address[14]. For each address, you can also view the history of all transactions made by that address along with dates.

Each block also identifies the username of the miner (the one who successfully solved the mathematical problem before everyone else) as well. This miner would have received a reward for doing this as well. This keeps miners motivated and therefore the system has an in-built reward within it. Each block is about 1.2 Mb. The block height or the number of the block is in the 6,00,000 series. All these transactions would roughly add up to 72 Gb of data. Since all the transactions on each block are hashed, the entire chain, without the record of the transactions would occupy less than 1 Mb of space. This allows the system to be scalable.

Everything about bitcoin transactions is public. The identities of the sender and the receiver are encrypted. Each wallet (of sender and receiver) has multiple unique hashed identities. This ensures that no one can trace previous transactions to one person.

The same is true for all crypto currencies. The famous ones with higher volumes have their ledger accessible from multiple websites.

4.7.3 Ethereum Blockchain

Ethereum has been a successful blockchain in terms of diversifying beyond a currency. At time of writing it is considered to be the default platform for coding DAPPS (stands for Distributed Applications). Solidity is the platform where smart contracts can be coded on to the Ethereum blockchain. Learning Solidity has been made easy by CrytoZombies.io. This platform offers a gamified system of learning how to code for Ethereum. The course is free and has levels that one progresses through in the journey of dealing with zombies.

4.7.4 Generate a Hash

There are a number of hashing algorithms. They have varied complexity. Some have a private key that can be changed for each transaction.

If you are familiar with the statistics software called R, then you can try out *digest* package to generate hashes. You can use the following code to generate the hash for any content *hmac ("PrivateKey", "Content", algo = "sha256").*

You can change the Private Key to anything you want. The content can be whatever you want. The algorithm can be specified to any out of sha1, sha256, sha512, crc32, mda5. These would all generate a different hash of the same content but would be internally consistent.

Alternatively, you can visitwww.sha1-online.com/ to hash anything you type.

You can hash images as well. Right click any image file and click on Open With... and click Notepad or any text editor. You will see the textual representation of the image. It is this text that would be hashed. Try editing this image in Paint or any other picture editing software. You could add a small dot, that might even be imperceptible to the naked eye. If you compare the text representation of this edited picture to that of the unedited it, you would be able to see the difference. Since the source text is different, the hash would also be different.

NOTES

1 DOA: October 10, 2019 https://dea.gov.in/sites/default/files/Report%20of%20the%20Steering%20Committee%20on%20Fintech.pdf
2 https://en.wikipedia.org/wiki/Punjab National Bank Scam
3 India Meteorological Department (IMD)
4 www.forbes.com/sites/mariyayao/2017/04/14/your-electronic-medical-recordscan-be-worth-1000-to-hackers/#5e0e1b7250cf
5 Usually above INR 20k in most cases
6 Source www.cbinsights.com/research/what-is-blockchain-technology/
7 www.livemint.com/Money/o4bSQ6CiUfjCIWDFDyZjnJ/Cryptocurrency-notlegal-tender-in-India-but-blockchain-get.html
8 https://dea.gov.in/sites/default/files/Approved%20Press%20Release%20on%20the%20Report%20and%20Bill%20
9 www.ey.com/en gl/global-review/2018/restoring-trust-in-the-wine-industry
10 Reserve Bank of India
11 www.thehindubusinessline.com/money-and-banking/indias-first-blockchainimplementation-goes-live/article23422835.ece
12 https://qz.com/947064/sweden-is-turning-a-blockchain-powered-land-registry-into-areality/
13 Rajasthan as accessed at eodb.dipp.in
14 These are encrypted.

5 Object Detection System with Image and Speech Recognition

Chung Van Le[1], *Vikram Puri*[1], *and Sandeep Singh Jagdev*[2]
[1]Duy Tan University, Vietnam
[2]Ellen Technology Pvt. Ltd., India

CONTENTS

5.1 INTRODUCTION

Nowadays, robotics have made human life very convenient, not only in industrial applications, but also in the research area of education, medical science and entertainment [1]. Many companies— namely Boston Dynamics—and researchers have worked to develop a number of robots to fulfill various requirements according to their research area and also to create a fruitful synergy of human-robot interaction. In addition, robots can easily deal with complex tasks that are difficult to deal with for humans. At present, numerous robots have captured the market but the robotic arm remains one of the most successful [2]. Usage of the robotic arm is the most important tool in factories for assembly processes, especially car assembly, and big manufacturing machines. In order to control coordination and movement of the robotic arm, accuracy, stability, and precision play an important role. Figure 5.1 represents robotics connected with other related technologies.

Object-recognition technology has rapidly increased in many domains such as object detection in 3D and 2D images, movement detection and with the handshaking of artificial intelligence (AI), these robots behaving like humans are called humanoid robots (HR). Moreover, HR also learns from its mistakes and improvises skills with

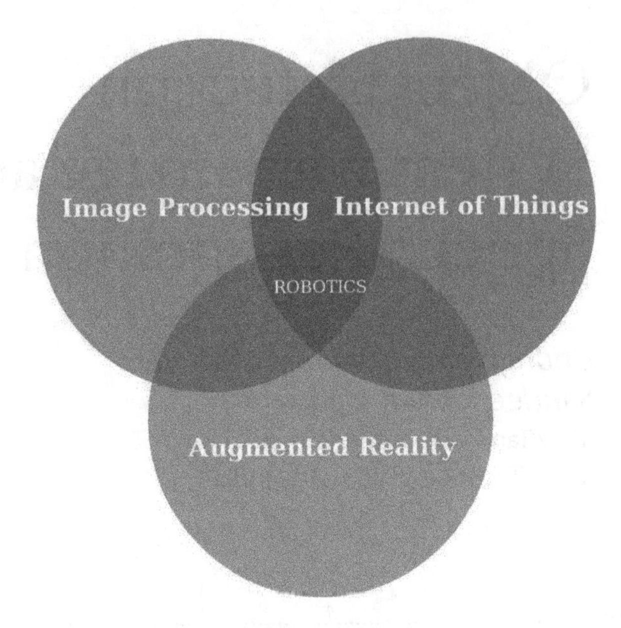

FIGURE 5.1 Robotics connected with other technologies.

the use of reinforcement learning. These days robots provide a helping hand to humans in their day-to-day life.

Vision is one of the intelligent core technologies in robotics. The research area in vision called "Computer Vision" is now considered from a scientific point of view for investigating how artificial computer vision can make a robot humanlike and what algorithm underlies it [3].

Open-Source Computer Vision (OpenCV) is a real-time library program developed by Gary Bradsky in 1999 [4]. It is an open-source library for both educational and commercial purposes. It supports C, C++ and Python interfaces and optimizes nearly 2,500 algorithms [5]. OpenCV plays a supportive role in the development of Computer Vision into a new futuristic world and enables millions of people to enhance their limits in productive work.

Many researchers have proposed and developed the robotic arm and visual system over the last few decades. Furuta [6] proposed a method to control trajectory tracking using a sensor-based feedback system. With the use of a laser beam, the proposed algorithm is used to achieve the desired coordinates for the robotic arm. Manasinghe [7] proposed an algorithm for the industrial robotic arm to contour problem cartesian velocity and joint torque. In this work, simulation is established to compute the coordinates of each joint in the robotic arm. Koga [8] developed a virtual model for the robotic arm to calculate joint coordinates when it picks up and places an object. Efe [9] presented a scheme to adjust the robotic arm fuzzy sliding controller with the use of the adaptive neuro-fuzzy inference system (ANFIS). Wang [10] proposed a robotic arm that is fixed on the mobile robot to detect signs or numbers. Image processing and detection are carried out through the use of a microcamera fitted on the robotic arm. Juang [11] developed a robotic arm system to grab objects via visual recognition. This system is equipped with two webcams: One webcam is employed to catch commands on screen and the other is used for word recognition. The authors [12] developed an application named APP that is integrated with the robotic arm as well as Raspberry Pi for the computation of Convolutional Neural Networks. Moreover, the camera is fitted on the robotic arm for the selection of donuts that matches with the customer flavor. Gaussian mixture is also part of the proposed system to segment the foreground and background picture. Karke [13] discussed real-time implementation of deep learning models with robotics application. Arduino Uno detects objects and classified them according to their category. Convolution Neural Network is used for further processing. Kumbhar [14] presented a project based on a low-cost robotic arm enabled with camera vision. Arduino Mega 2650 is deployed as the main controller as well as six different motors controlled via controller. Object detection and image edge detection has been recognized via Raspberry Pi.

The main concept of this chapter is to give vision capability to the robotic arm through the use of image processing and speech recognition. An ultrasonic sensor is also integrated with the camera on the robotic arm to check and calculate the distance between the object and the robotic arm. The proposed study is discussed as follows: Section 5.2 explains the methodology, Section 5.3 illustrates the results and discussion and Section 5.4 concludes the proposed study.

5.2 METHODOLOGY

5.2.1 Architecture

Our proposed work is categorized into different sections; namely speech recognition, image capturing, processing, sensing the distance of the object, flow of algorithm and robotic arm functions.

Raspberry Pi serves as the backbone of our proposed system. Image processing is done through the use of a camera and OpenCV. Raspbian operating system(OS) [15] used in the Raspberry Pi which is based on Debian operating system provides over

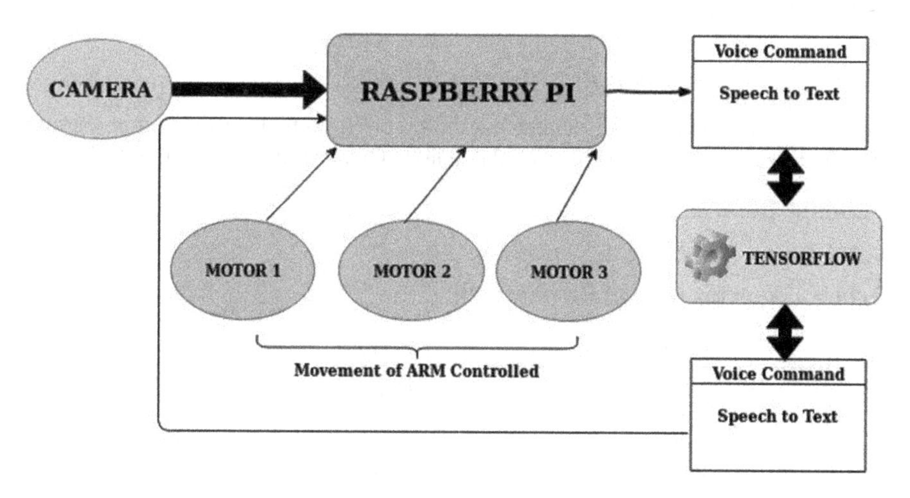

FIGURE 5.2 Proposed System Architecture.

35,000 pre-installed packages and pre-compiled software such as python, sonic-pi, and java. In addition, it is more than a pure OS. OpenCV, installed through Linux commands, provides library packages to process images taken through the camera installed with Raspberry Pi. These libraries are based on python. Two different sensors, namely ultrasonic and IR sensor, check the distance and location of the object. The sensor works on the principle of calculating the distance of the reflected wave. The formula for distance calculation of the reflected wave is:

$$\mathbf{D = \tfrac{1}{2} T * SS} \tag{1}$$

Where D is Distance, T is time and SS is the speed of sound. SS varies with humidity and temperature.

For the robotic arm, Servo motors are employed. These motors work on the pulse width modulation (PWM) principle and control through three wires, 1) Power 2) Ground 3) Signal. In PWM, there are three different pulses in which the motor will work; minimum pulse, maximum pulse and the last one is repetition pulse and it rotates around 180 degrees (both sides 90 degrees). PWM pulse decides the position of the motor shaft stamped with time duration. In the proposed work, three servo motors are employed inside the robotic arm, one for up and down and another one is to grasp the object. The last one is for rotating the arm.

5.2.2 Circuit Diagram

In the proposed study, Raspberry Pi plays a major role in controlling the robotic arm and capturing images while processing according to the requirement. Raspberry Pi is a small credit card based computer operated on Debian based OS. Table 5.1 represents the technical specifications of the Raspberry Pi.

The camera is connected to Raspberry Pi through the USB port. IR sensor is based on three pins: 1) power 2) ground 3) output and connected to Raspberry Pi general-purpose input-output (GPIO) pin. Three servo motors are also connected to GPIO PWM pins(see Figure 5.3).

5.2.3 Speech Recognition

Speech recognition[16][17] is a language-based program that is used to input human speech, decrypt it and change into readable text. This technique helps to filter words,

TABLE 5.1
Technical Specification of Raspberry Pi

S.No	On-Chip	Feature
1.	CPU	Quad Core Arm Cortex
2.	RAM	2 GB
3.	Storage	MicroSDHC card
4.	Power	5Volt – 2 ampere
5.	Graphics	Broadcom VideoCore

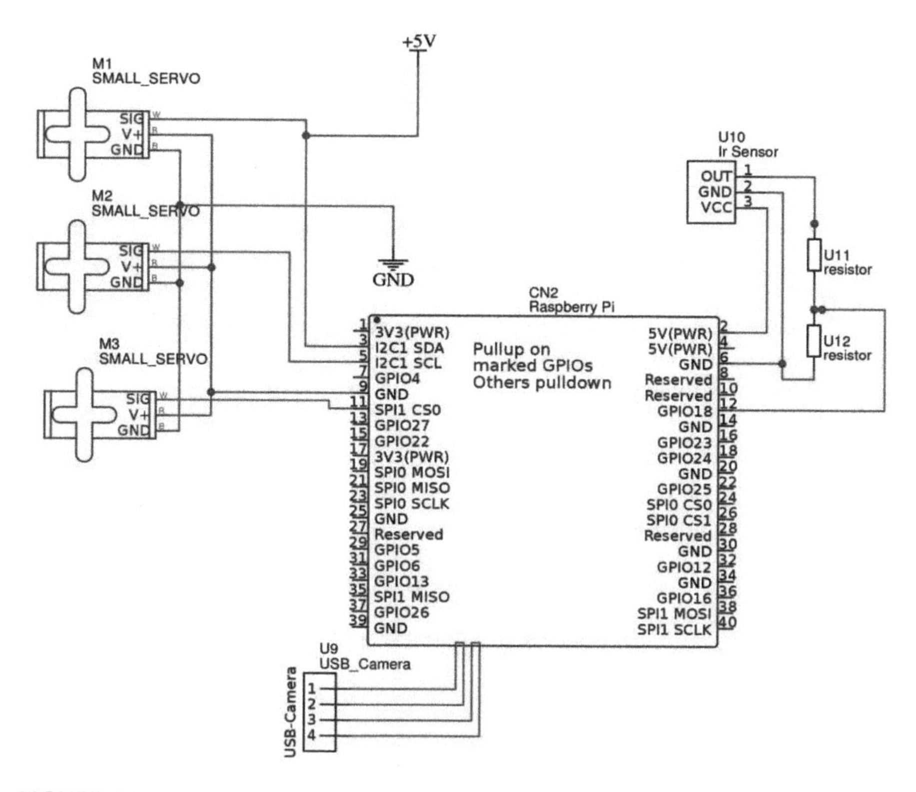

FIGURE 5.3 Circuit Diagram of Proposed work.

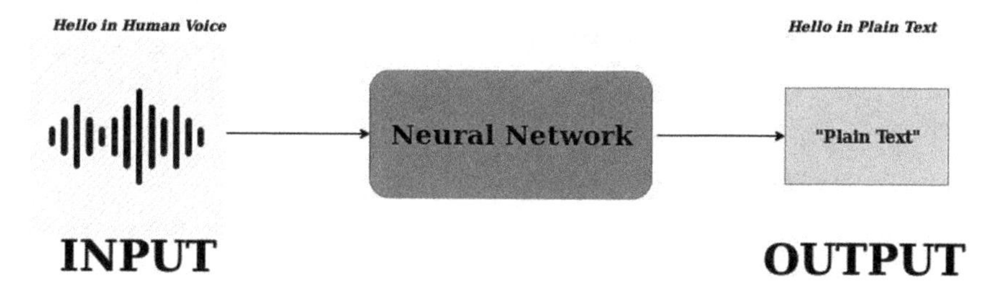

FIGURE 5.4 General Speech Recognition Process.

change them into digitize format, and analyze that sound. Figure 5.4 represents the general speech recognition system.
Simple Speech recognition module is categorized into three parts as follows:

1. **Input: Human:** Voice is used as source of input for this module.
2. **Neural Network:** Natural Language Processing (NLP) and Neural Network (NN) break speech into a number of components that can help to easily interpret it. After conversion, these components change into a digital state and

analyze these states. NN is used to train the dataset of specific words and phrases and create predictions for the new voice related to input data.
3. **Output:** In the last stage, it transcribes the input voice into text format.

5.2.4 Tensor Flow

Tensor flow [18] is an open-source platform especially designed for machine learning and deep learning that helps to provide end-to-end solutions. It provides a huge number of tools and flexible platforms that motivate and push researchers to extend their creativity limit. Tensor flow gives stable Python and C++ APIs. Moreover, these APIs are compatible with other languages in unstable versions. The main features of Tensor flow are as follows:

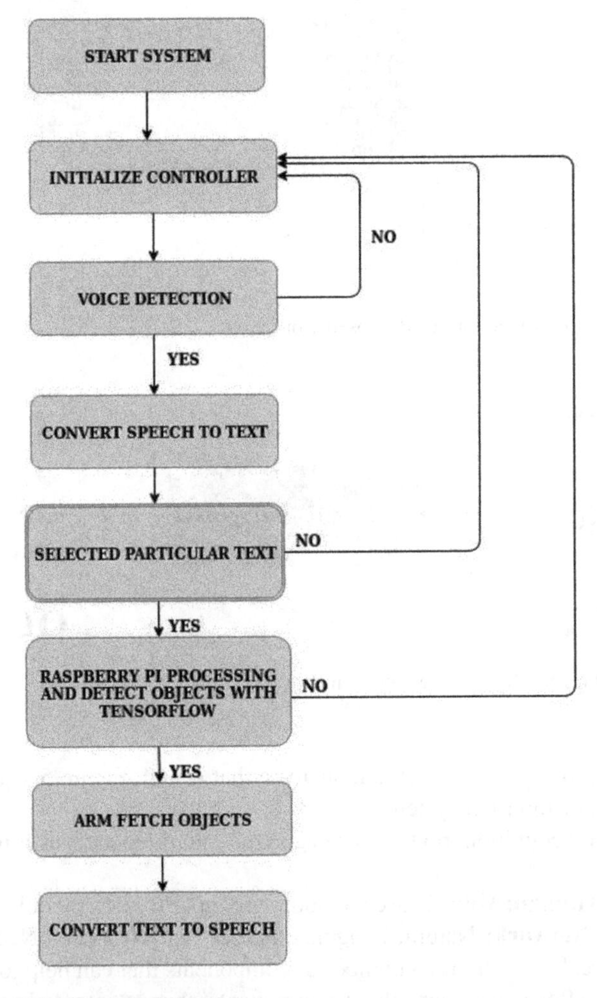

ALGORITHM 5.1 Proposed Workflow.

1. **Model Building:** Training and Building models in machine learning is quite easy compared to other system models. With the introduction of Keras, this helps to improve model debugging and model iteration.
2. **Robust:** Due to the power of clouds, it is easy to train and train deploy models anywhere.
3. **Research Experimentation:** This provides a flexible architecture that helps to implement from idea to practical implementation. Especially regarding results, it provides these in the form of graphs and statistical form as well.

5.2.5 Proposed Algorithm

Algorithm 1 presents a proposed workflow. In this system, the arm is activated through the voice and if there is no matching voice, it rechecks. In the next step, voice is converted into plain text, processing this text in the TensorFlow and checking the libraries and detecting the object with the convolutional network. The arm is synchronized with the TensorFlow. A camera captures the image which is processed through the use of installed OpenCV in the Raspberry Pi. Sensors check the location of the object. If the object is within range of the robotic arm it will grasp it; otherwise the robotic arm will not move. With the use of text to speech, it can convert into voice instructions such as object name, specification etc.

5.3 RESULTS AND DISCUSSION

In this section, the input signal (see Figure 5.5) is used to recognize the particular object. For image recognition, Tensor flow is deployed and measured through the accuracy score of the model. The accuracy of the training dataset is 0.697.

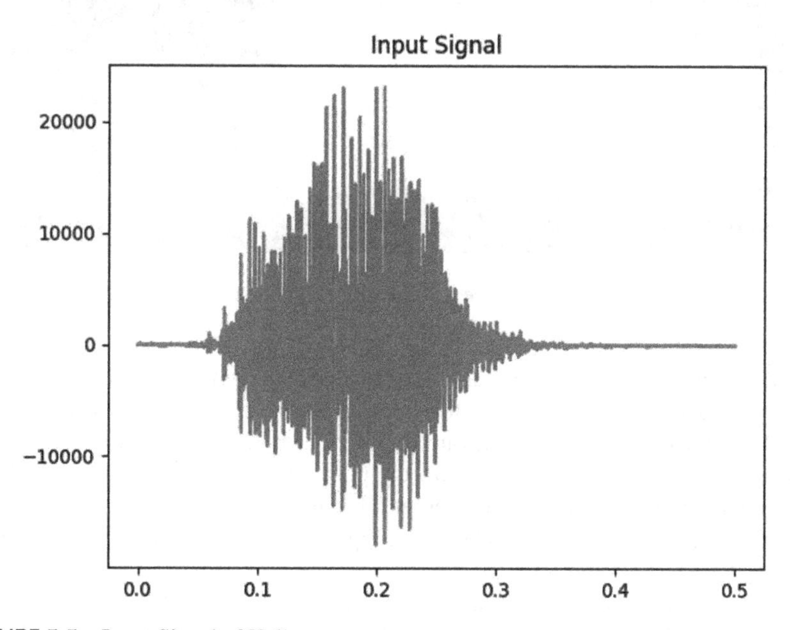

FIGURE 5.5 Input Signal of Hello.

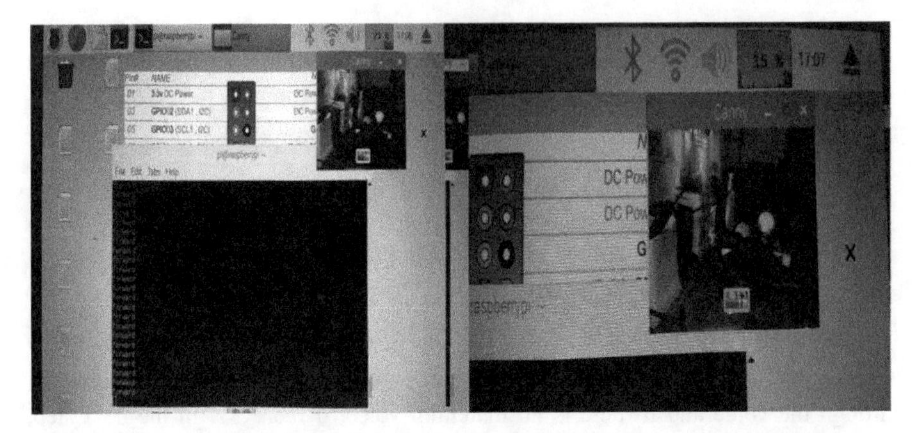

FIGURE 5.6 Left Side: Direction of Robotic Arm; Right Side: Detection of Object.

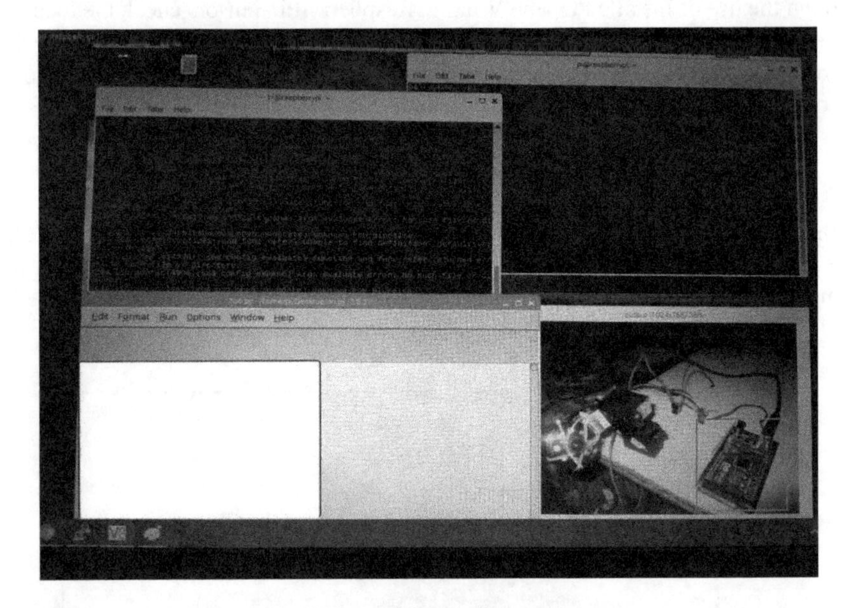

FIGURE 5.7 Activated mode of Voice and Image Recognition.

In this section, Figure 5.6 and Figure 5.7 represent the outcome of the results from our proposed work. In Figure 5.6, the terminal shows the direction of the robotic arm which moves forward or backward. Figure 5.6 shows the detection of an object with a yellow screen. The most noteworthy point of this proposed system is that it works remotely through use of the internet. Through the use of SSH, the robotic arm captures images and grasps crucial objects. Figure 5.7 presents an activated mode of image and speech recognition.

Figure 5.8 represents the system proposed to fetch objects with a camera mounted on the arm. With the aid of speech recognition, it is activated through voice. Below

FIGURE 5.8 Robotic arm enabled with Camera.

the arm, IR and ultrasonic sensors are mounted that help camera to locate the object location as well as maintain an exact distance between the object and arm.

5.4 CONCLUSION

In this chapter, a vision- and voice-based robotic arm is proposed. This proposed system provides an opportunity to tackle real-world problems such as remote surveillance; tasks impossible for humans. The integration of day-to-day life problems with complex visions can improve image processing research. OpenCV libraries installed on the Raspberry Pi allow focus on image processing with minimal labor. In this study, the robotic arm is equipped with a camera, ultrasonic sensor and IR sensor that increases the accuracy of selection of objects from exact coordinates. Object detection and location extraction techniques are executed with the aid of image processing methods; namely object extraction techniques, matching pre-installed templates.

REFERENCES

1. Juang, J. G., Tsai, Y. J., & Fan, Y. W. (2015). Visual recognition and its application to robot arm control. *Applied Sciences*, *5*(4), 851–880.
2. Manigpan, S. (2010). *A Simulation of 6R Industrial Articulated Robot Arm Using Neural Network* (Doctoral dissertation, University of the Thai Chamber of Commerce).
3. Ejiri, M. (2007, November). Machine vision in early days: Japan's pioneering contributions. In *Asian Conference on Computer Vision* (pp. 35–53). Springer, Berlin, Heidelberg.
4. OpenCV Introduction: https://opencv.org/ (Accessed on 16/11/2019).
5. Bradski, G., & Kaehler, A. (2008). *Learning OpenCV: Computer vision with the OpenCV library*. "O'Reilly Media, Inc.".

6. Furuta, K. A. T. S. U. H. I. S. A., Kosuge, K. A. Z. U. H. I. R. O., & Mukai, N. O. B. U. H. I. K. O. (1988). Control of articulated robot arm with sensory feedback: Laser beam tracking system. *IEEE Transactions on Industrial Electronics*, *35*(1), 31–39.
7. Munasinghe, S. R., Nakamura, M., Goto, S., & Kyura, N. (2001). Optimum contouring of industrial robot arms under assigned velocity and torque constraints. *IEEE Transactions on Systems, Man, and Cybernetics, Part C (Applications and Reviews)*, *31*(2), 159–167.
8. Koga, M., Kosuge, K., Furuta, K., & Nosaki, K. (1992). Coordinated motion control of robot arms based on the virtual internal model. *IEEE Transactions on Robotics and Automation*, *8*(1), 77–85.
9. Efe, M. Ö. (2008). Fractional fuzzy adaptive sliding-mode control of a 2-DOF direct-drive robot arm. *IEEE Transactions on Systems, Man, and Cybernetics, Part B (Cybernetics)*, *38*(6), 1561–1570.
10. Wang, W. J., Huang, C. H., Lai, I. H., & Chen, H. C. (2010, August). A robot arm for pushing elevator buttons. In Proceedings of SICE Annual Conference 2010 (pp. 1844–1848). IEEE.
11. Juang, J. G., Tsai, Y. J., & Fan, Y. W. (2015). Visual recognition and its application to robot arm control. *Applied Sciences*, *5*(4), 851–880.
12. Chen, O. T. C., Zhang, Y. C., Lin, Z. K., Kuo, P. I., & Lee, Y. L. (2019, August). Camera-in-Hand Robotic Arm Using a Deep Neural Network to Realize Unmanned Store Service. In *2019 IEEE Intl Conf on Dependable, Autonomic and Secure Computing, Intl Conf on Pervasive Intelligence and Computing, Intl Conf on Cloud and Big Data Computing, Intl Conf on Cyber Science and Technology Congress (DASC/PiCom/CBDCom/CyberSciTech)* (pp. 833–839). IEEE.
13. Kakde, Y., Bothe, N., & Paul, A. (2019). Real Life Implementation of Object Detection and Classification Using Deep Learning and Robotic Arm. *Available at SSRN 3372199.*
14. Kumbhar, S., Mathurekar, D., & Lobo, D. (2019). *Robotic Arm with Vision* (Doctoral dissertation).
15. Raspbian Operating System. www.raspbian.org/ (Accessed on 17/11/2019)
16. Graves, A., Mohamed, A. R., & Hinton, G. (2013, May). Speech recognition with deep recurrent neural networks. In *2013 IEEE international conference on acoustics, speech and signal processing* (pp. 6645–6649). IEEE.
17. Hinton, G., Deng, L., Yu, D., Dahl, G., Mohamed, A. R., Jaitly, N., ... & Sainath, T. (2012). Deep neural networks for acoustic modeling in speech recognition. *IEEE Signal processing magazine*, *29*.
18. TensorFlow: www.tensorflow.org/ (Accessed on 10/12/2019)

6 Blockchain Applications and Implementation

Deepak Kumar Sharma, Tushar Pardhe, Yash Kulshreshtha, and Shivani Singh
Department of Information Technology,
Netaji Subhas University of Technology,
New Delhi, India

CONTENTS

6.1 INTRODUCTION

Recently, there has been a lot of buzz around bitcoin and other cryptocurrencies. But the driving technology for bitcoin i.e., blockchain, is often misconceived and is less explored. Blockchain has been one of the buzzwords for a few years now. These days, blockchain is everywhere. You cannot read the news without coming across news related to blockchain, bitcoin, cryptocurrencies, etc. It has been said that blockchain will do for transactions what the internet has done for information. Experts claim that blockchain will touch all vital fields and has the ability to alter payments, economics, healthcare and even politics around the world. Buying a house or registering a vehicle can prove to be quite a painful process due to the number of intermediaries one has to go through. The primary reason for the popularity of this technology is that it allows increased transparency and efficiency in the exchange of almost anything. It reduces cost and promotes trust among parties. To sum up, blockchain has the ability to change how the world works and it has the potential to be called the most groundbreaking technology of this century.

Despite being such a revolutionary technology, many people still do not have a clear idea about what it is and how it works. The true potential of this technology lies in awareness of it. So this is precisely the agenda of this chapter i.e., what exactly is blockchain and why does it seem to have such limitless applications? This chapter explains blockchain's significance and implications along with an outline of a brief history behind it, starting with Section 1.1.1 which gives a general idea about blockchain. Then we shall try to understand the origin of blockchain in Section 1.1.2. Further, Section 1.1.3 briefly discusses its implementation.

6.1.1 What is Blockchain?

Blockchain is a way of recording transactional data or tracking assets. Assets can be tangible or intangible. Practically, anything of value can be tracked using a blockchain network with lesser costs and improved transparency. Blockchain is a database distributed across a network, where the nodes of the network are in sync with each other. All the nodes have the exact same copy of the blockchain. There is no single entity or authority which is in charge of controlling the database. Any modification to this database is done after broad agreement from the participants of the network. This database is append-only i.e. data can only be added to the database, and only after agreement from all participating entities [1].

With reference to past events, when various people had to depend on the same data, they relied on a trusted third party to control the origin of data. But with the emergence of blockchain, these groups of entities can agree on events without the requirement of a third party.

Blockchain is basically a decentralized ledger, building a linked list of records, called blocks, across a point-to-point network where every block contains a certain number of transactions. The order of the blocks in the chain specifies the order in which the transactions take place. Each block consists of:

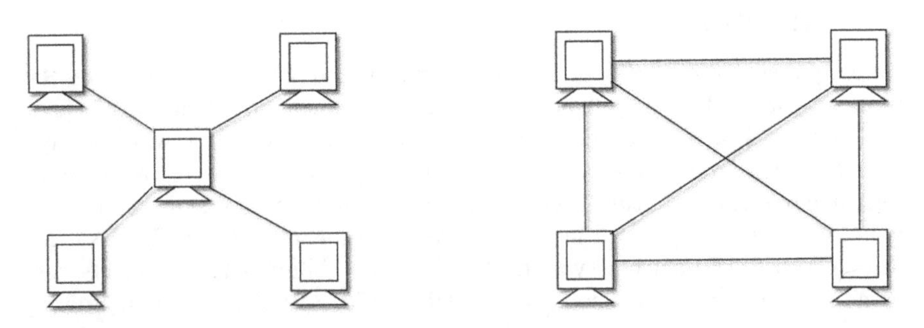

FIGURE 6.1 Centralized versus Decentralized systems.

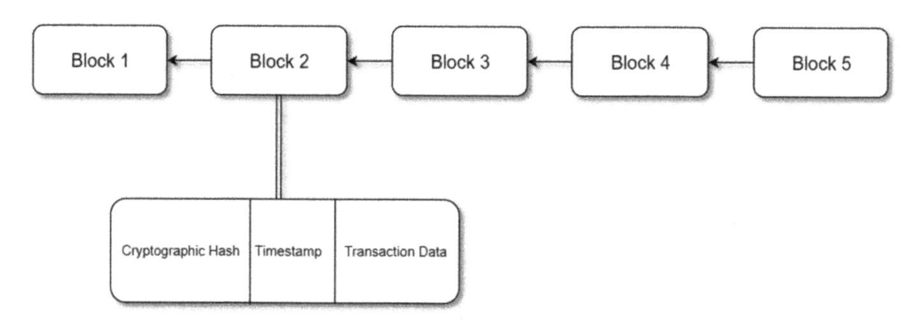

FIGURE 6.2 Structure of a Block.

1. Cryptographic hash of the previous block
2. The timestamp
3. Transactional data

In other words, the blockchain functions as an open, shared, trusted public chain that maintains a continuously growing list of transaction data between two entities efficiently and in a verifiable and persisting manner.

Due to the structure of each block, blockchain is resistant to modification of data. If someone tries to change the transaction history in some block, then all the blocks from that block until the current block will have to be modified and this will have to be reflected in every duplicate of the blockchain on the network. Since the hash in the next block is a function of the contents of this block, the next block will also have to be changed. This change will result in a different block hash. The same process will have to be circulated to the latest block in the chain. Maintaining the blockchain becomes extremely difficult once it is modified. The computing power required to accomplish this is colossal. But these days we have computers with very high processing powers and for such computers, calculating hundreds of thousands of hashes is a matter of a few seconds. So, to avoid tampering with data, blockchains use a technique called *Proof of Work (PoW)* which requires miners (nodes in the network) to solve a computational problem. The time and effort required to solve the problem is very high as compared to the time required to verify the result of the problem. Once a miner finds a solution to a block, it broadcasts the block to the network. All the other nodes verify this solution and the block is added to the chain. PoW has become a widely used consensus algorithm and is used by many cryptocurrencies [2].

Now let's try to understand the working of the blockchain with the assistance of an example:

Suppose a person X wants to give ₹10,000 to Y.

Without blockchain, X would send a request to his bank to initiate a transaction of ₹10,000 from his account to Y's account. This request will contain the sender's and receiver's account details along with the transaction amount. The bank would check a few things such as if X actually has ₹10,000 in his account. If everything works out fine, the bank would transfer ₹10,000 from X's account to Y's account.

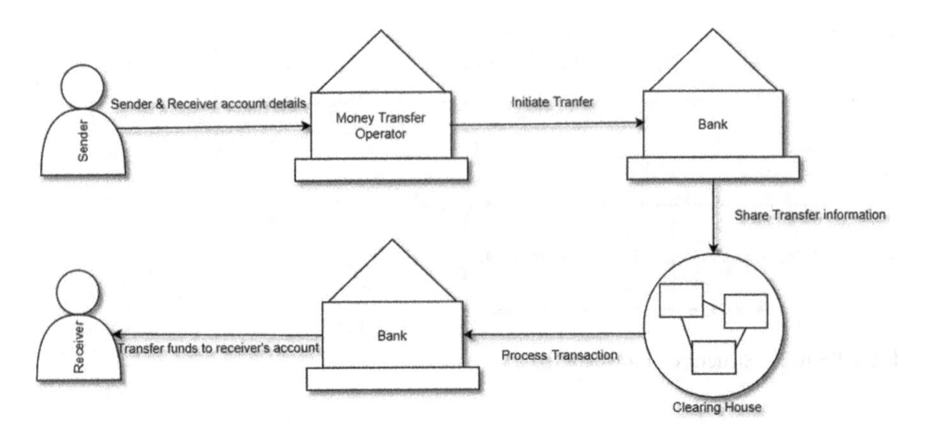

FIGURE 6.3 Transaction without blockchain.

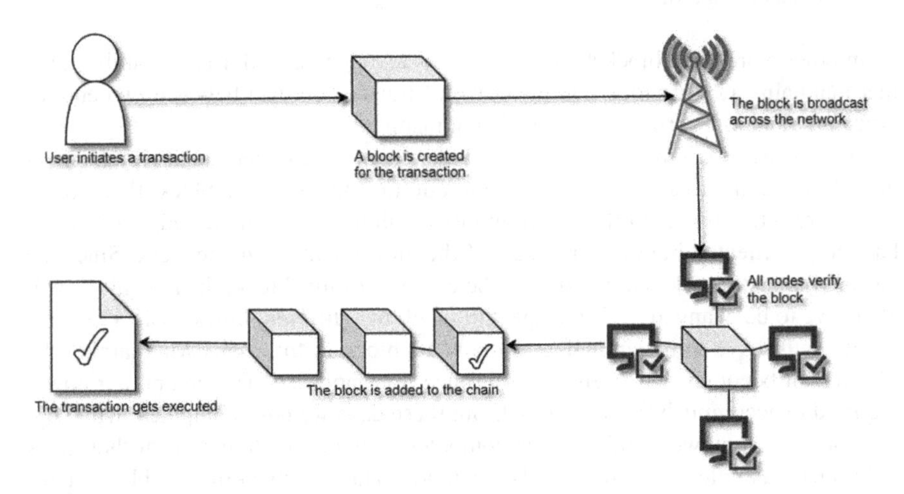

FIGURE 6.4 Transaction with Blockchain.

With blockchain, X creates a transaction of ₹10,000 to Y. This transaction is included in a block and sent over the internet. All the nodes in the network check whether this is a valid transaction. If it is valid, this block is added to the chain and the transaction gets executed i.e., Y has X's ₹10,000.

Blockchain is by all accounts the driving innovation behind the next-generation internet or Decentralized Web and gives an answer to the age-old human issue of trust. Now, let us look at the key attributes of blockchain which justify the above statement and such endless applications of blockchain:

1. Removing the role of a central governing authority: Blockchains are managed by a group of nodes forming a point-to-point network rather than some intermediary. Instead of being verified by a third party, a transaction is verified by

every node in a network. This has several benefits like reduced transfer costs and improved transparency.

2. Distributed: Every node has a copy of the blockchain. So, if someone tries to tamper with the data, he/she will have to modify data in all the copies of blockchain stored at every node in the blockchain network. This prevents all of the information from being tampered with at the same time. Unless hackers can tamper with the transaction records on millions of nodes, they will not be able to successfully falsify the blockchain data.
3. Immutable: Any transaction data, once added to the blockchain cannot be modified. Data can only be added to the blockchain in a time-sequential order. It is considered practically impossible to change any data.
4. Consensus: This is the attribute which gives blockchain the power of decentralization. The ledger is updated by consensus. No central authority is in charge of making changes to the ledger. Any updates made to the blockchain are verified against strict conditions and added to the blockchain only after a consensus has been reached among the nodes in the network.
5. Security and reliability: There is no single point of failure in the blockchain system due to decentralization. There is no part of the system that has the ability to cease functioning of the entire system, if it fails. Also, there are no weak points in the network from where information can be tampered with. This avoids malicious attacks, thus improving overall reliability of the blockchain. Furthermore, transactions are digitally signed and encrypted, which ensures high security.

6.1.2 History of Blockchain

For a better understanding of blockchain, we must trace blockchain's origin. We should try to understand the context in which it was developed to perceive it better. Since the 'big 4' are investing in it, blockchain has gained even more importance. Optimists claim that blockchain is going to be of more prominence in the future. So, let us see what journey blockchain had before it became a matter of interest to numerous people around the world.

In 1991, Stuart Haber and W. Scott Stornetta devised a method to cryptographically secure a chain of blocks to develop a system where time-stamped documents cannot be tampered with. In 1992, they tried to incorporate Merkle trees into this system to make it more efficient so that multiple documents could be stored in one block. However, it was in 2008 that blockchain finally began to gain importance [3].

In 2008, blockchain and bitcoin were conceptualized by a person (or a group of persons) named Satoshi Nakamoto. Satoshi Nakamoto has chosen to remain anonymous till now. In Nakamoto's paper 'Bitcoin: A Peer-to-Peer Electronic Cash System', he introduced bitcoin to the world and explained its underlying principles and how it works. In this paper, he incorporated many concepts such as cryptography, networks, calculus, etc. He tried to devise an electronic cash system based on cryptographic proof. In the following years, bitcoin became popular and the underlying concept i.e., blockchain, became even more popular. Since then blockchain has evolved and found its place in many applications beyond cryptocurrencies [3].

6.1.3 How it is Developed/Implementation

Blockchain was introduced with the intention of disrupting the financial sector. Many banks and financial institutions have taken advantage of this technology to make transactions more secure. But now this technology is not confined just to payments and economics. Every industry, ranging from healthcare to manufacturing, from travel to retail, is investing in it.

The underlying concept behind blockchain is similar to that of a database but the way this database is handled is entirely different. For developers willing to learn to develop blockchain, it is important to understand some key concepts which form the crux of this technology:

1. Decentralized consensus: Blockchain is a decentralized peer-to-peer system. There is no mediator among the nodes in the network to govern the exchange of information. No involvement of a central authority keeps the system free from corruption. So, to make a decision in a blockchain, the participants of the network need to come to a consensus via a consensus algorithm.
2. Smart contracts: This is a computer protocol that controls the transfer of assets between entities. It defines the terms and conditions of a transaction. It works in the same way as a traditional contract. It is a piece of code that requires an exact sequence of actions to take place to facilitate an agreement between the entities involved.
3. Mining: Mining is the process of verifying a transaction and adding it to the chain of blocks. It basically involves running the consensus algorithm along with creating a hash of the previous block which is not easy to tamper with. The main purpose of mining is to add transactions to the blockchain in such a way that it becomes practically impossible to modify them in the future.

The basic concepts alone are not enough to develop a blockchain network. You should also be familiar with object-oriented programming languages like C++, Java, Python, Solidity, etc. You should also have some knowledge about data structures such as linked lists, hash tables, associative arrays, etc. that play an important part in creating the structure of blocks in a blockchain network. Some understanding of cryptographic techniques such as secure hashing algorithm (SHA) is also required which is useful while creating hashes for blocks [4].

The security and verifiability of transactions in the blockchain network is possible only because of the presence of consensus algorithms which play a vital role in the blockchain network. Consensus algorithm is a mechanism via which all the participants of the blockchain network reach an agreement. So now let us discuss some of the most common consensus algorithms used for blockchain development:

1. Proof of Work (PoW): This is the first consensus algorithm used in the blockchain network. It is used by the majority of cryptocurrencies. It involves

solving a complex computational problem. The principle of this technique lies in the fact that the time and effort required to solve this problem is much greater than the resources required to verify the result of the problem. The node which solves this problem first gets to mine the next block. When a miner wants to add a transaction to the blockchain, he/she does the PoW for the corresponding block and broadcasts it to the blockchain network. All the participating entities of the network verify and validate this block and the block gets added to the blockchain. Whenever a new block is mined, the miner gets rewarded with some currency [3][4].
2. Proof of Elapsed Time: PoET is one of the best and fairest consensus algorithms that is used with a permissioned blockchain network to find out the next miner. In this technique, every node is equally likely to be the next miner. All participating entities are required to wait for a random amount of time. This random wait time is generated by every node. The node which wakes up first after being in sleep mode (i.e., shortest wait time) for this designated time is chosen as the next miner. This miner add a new block to the blockchain and transmits the necessary information across the network.
3. Proof of Stake: This algorithm is an alternative to PoW. Instead of checking which node has more computational power to solve a problem, it works on stake. Stake here is the amount of currency a miner is willing to lock up for a certain amount of time. All the nodes of the network verify the blocks by placing a bet on it if they think that a certain block should be added to the chain. The nodes get rewards proportionate to their bets and their stakes increase accordingly. The node with the highest stake is chosen to mine the next block [2].

Now let us have a brief introduction of some tools useful in the development process of blockchain.

1. Geth/Parity: To interact with the blockchain network, we require a node capable of establishing a point-to-point connection with other nodes of the blockchain network. Geth and Parity are interfaces used to create a full implementation of an Ethereum node. The only difference is that Geth uses Go and Parity uses Rust for node implementation.
2. Mist: This is a program which acts as a wallet to store and send ether (the currency used on the Ethereum blockchain).Whenever you want to send or receive ether or invest on the Ethereum blockchain, you will require a wallet to carry out transactions and that is where Mist is helpful.
3. Blockchain Testnet: This is a platform used to test a decentralized application before deploying it to the main network. It provides dummy currency having no value to carry out transactions.
4. Solidity Compiler: Solidity is a loosely typed programming language used to write smart contracts on the Ethereum blockchain network. The Solidity compiler converts the solidity code into a form readable by Ethereum Virtual Machine.
5. Remix: This is a Solidity-based compiler that allows users to develop smart contracts for the Ethereum blockchain. It allows us to debug, deploy and test smart contracts.

It becomes quite complicated to develop a blockchain network from scratch due to the complex concepts involved, such as cryptography, networks, hashing, etc. So nowadays we have platforms which allow us to create our own blockchain by defining the upper-level information such as terms of a transaction and not paying much heed to the lower-level implementation details. Some of these platforms are listed below:

1. Ethereum
2. Openchain
3. Hyperledger Fabric
4. Hyperledger Sawtooth
5. EOSIO
6. Multichain

6.2 ETHEREUM BLOCKCHAIN NETWORK

Up to this point we all have become familiar with blockchain which is essentially a distributed ledger. With the word blockchain, bitcoin comes to mind. Bitcoin is blockchain technology but it only deals with currency transactions. A much more versatile and agile technology is needed in order to make widespread use of blockchain. Before developing a blockchain-based application, one needs complex coding, cryptography and advanced applied mathematics experience. To address this issue a platform called Ethereum was developed. Ethereum is an environment or platform which provides integrated tools and protocols to develop general-purpose applications. Unlike bitcoin-blockchain, Ethereum blockchain allows transactions of not only currency but also other information and values using blockchain.

Ethereum transactions are based on smart contracts that are like normal contracts but also automatically binding. Several languages are developed to write smart contracts; one of the most famous is Solidity. The dynamic nature of Ethereum opens up vast possibilities for its applications. In the upcoming sections we will dig a bit deeper into Ethereum and will have a look at technologies associated with it.

6.2.1 What is Ethereum?

When we hear "Ethereum" we understand it to mean a cryptographic cash—like bitcoin. While that thought is not totally misguided—it is basic to grasp that Ethereum is a long way beyond an essential advanced cash—rather it is an open programming stage produced using blockchain development that engages creators and programming specialists to build and pass on the tremendous scope of decentralized applications.

Inside the Ethereum system, there is a cryptographic cash called ether. It is used to control applications built using the Ethereum blockchain. The purpose of Ethereum is to change how the web transitions, in light of the fact it empowers distributed systems to work without using any intermediary. Ethereum licenses programming applications run on an arrangement of various private PCs. This is generally called an appropriated system. Much equivalent to most applications, information is taken care of on a remote server, which is in a general sense just a remote PC with a consolidated

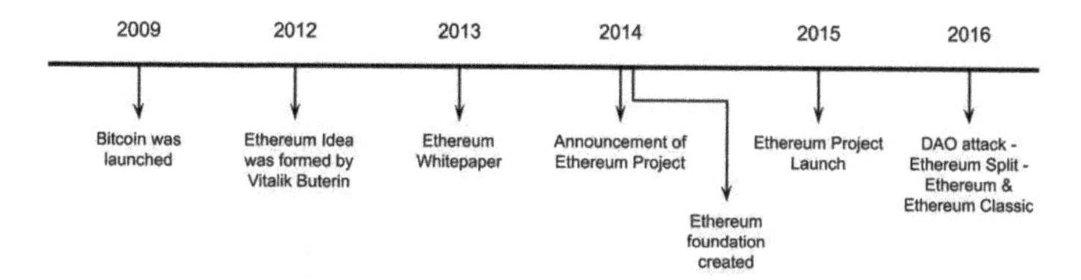

FIGURE 6.5 Ethereum Timeline.

database that contains the site's information and meta-information. If that server is harmed, all of the data in the structure and database and the site dissipate. With the use of blockchain advancement, that identical database and structure is appropriated among a tremendous number of people's PCs and systems, so all data in the database is open and the database and system cannot be shut down in any practical sense as long as various PCs are still adding to it [7].

In short this implies "brought together corporate uber PCs and cloud servers are replaced with a gigantic, decentralized arrangement of various little PCs that are used and maintained by volunteers (people like you and me) from all around the world."

In 2013, Vitalik Barutin, a computerized cash master and programming engineer, first released a paper on Ethereum. This progress was later financed by a web crowdfunding event which was held in mid-2014. On 30 July 2015, Ethereum was launched, with 720 lacs of ether being predefined over the framework and sorted out and set apart as "premed". This aggregate is around 68 percent of the most bought stock in 2019. A half-year later makers chose to keep Ethereum a non-benefit substance. Ethereum Foundation (Stiftung Ethereum) was likewise framed.

It was decided initially that Ethereum would use proof-of-work methodology; therefore its developers have to "mine" or validate nodes, which rewards them with ether tokens. But recently this methodology has faced worldwide friction due to its energy inefficiency. Environmentalists are criticizing it for its high energy usage. In recent developer meetings of Ethereum, a "**proof of stake**" model has been discussed. We will discuss these problems later in this chapter.

While developing Ethereum it was kept in mind that it is *Turing complete* and *stateful* which means that these smart contracts pick up where they left off. In other words Ethereum blockchain can remember various kind of data. Ethereum was kept open source. But these principles created some specific difficulties for the Ethereum blockchain. Because it is an open-source platform, anyone can build their own app on top of it; it is very important to increase usage of it and make it popular among developers but its downside is that it makes the whole network slower because each node has to work much longer. That being said, there are not many fully functioning blockchains to date that can simulate real decentralized applications. Moreover almost none is faster than Ethereum among currently existing blockchains (Hashgraph and Dag offer faster transactions per second).

DAO, a decentralized autonomous organization, raised a record 150 million in crowdsale to fund the project in 2016. In the same year an unknown hacker removed 50 million ether in June from DAO. A debate among the crypto community was

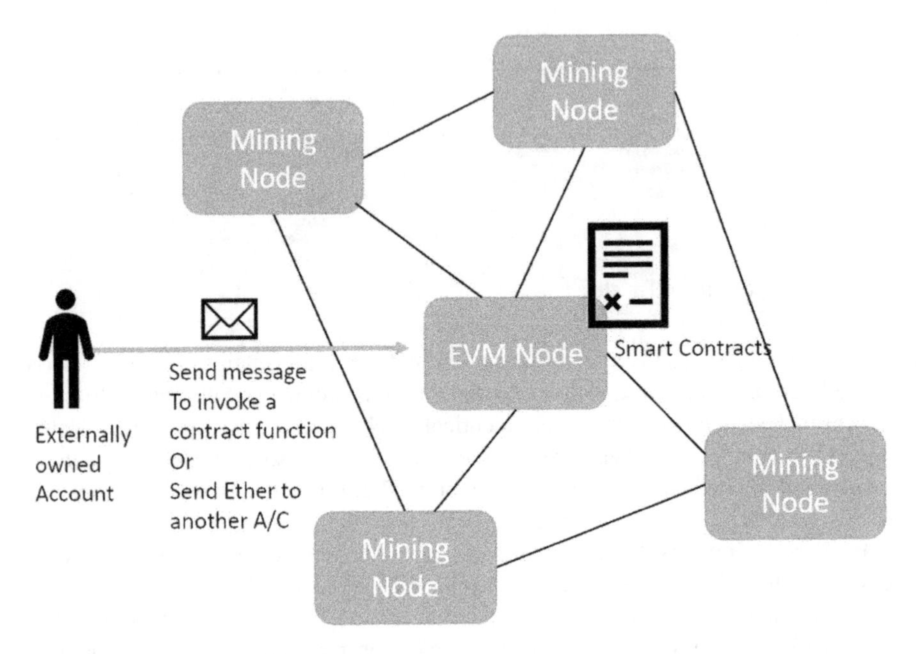

FIGURE 6.6 Ethereum Basic Architecture.

started by this incident over whether Ethereum will reuse the funds involved by creating a controversial "hard fork". Due to this whole controversy, the Ethereum network was split into two parts. Ethereum, which we are discussing, continued on the forked blockchain, while the original blockchain continued by the name Ethereum Classic. This Hard Fork strategy started a rivalry between these two blockchain networks. After a tough and critical situation involving DAO again, Ethereum had to be split twice in the last three months of 2016 to counter other attacks. After that Ethereum had to increase DDoS protection, bearing the blockchain and preventing spam attacks by hackers by the start of December 2016.

It is reasonable to view the Ethereum ecosystem as a large market, where a wide variety of goods and services are available and all transactions are handled through a power grid provided by ETH Currency (Ether). This means that for every activity on the Ethereum blockchain there is a small fee called "gas" (usually a few US cents at current prices). Whether we send ETH or tokens or interact with deals, games or services, we need a little ETH in our wallet. But we do not need ETH to get transfers. A publicly distributed general ledger is created for transactions by ether, which is a basic token for Ethereum's operation. It can be used to pay for transactions, a processing unit used in gas and other state transitions. Sometimes ether is called Ethereum, which is incorrect.

The ticket symbol for ether is ETH. It is also traded on cryptocurrency exchanges. For ether the currency symbol is usually the Greek capital letter Xi. It is also used to pay invoice services and transaction fees on the Ethereum blockchain network. Ethereum addresses a common hexadecimal identifier associated with 20 octets of the Kekak 256 hash (Big Endian) of the public ECDSA key (the curve used is called "secp256k1"). In this address two digits represent one byte, and it means that the addresses contain a total of 40 hexadecimal digits. For example: 0xc764Y7eA0ba3 2784cE839613nhaBA85786947268. Contract addresses also have this same format but they are determined by creation and sender transactions. The accounts of users cannot be separated from accounts of contracts for which only blockchain data and an address is specified. Any 256 hash entered into the format described is valid. Even if this does not correspond to the account with a private key or contract, it is valid, whereas bitcoin uses a base 58 check to make sure the addresses are entered correctly [8].

6.2.2 Is Ethereum the Future?

Ethereum blockchain is the second most popular blockchain in the cryptocurrency industry but it is not free from issues. Even though the second largest, it has some scalability issues. Scalability of any blockchain refers to the number of transactions a blockchain can handle per second. Ethereum can handle only 15 transactions per second. This is a problem many people faced in the early days of the project. But in the case of Ethereum, as it gains popularity, this problem is becoming more and more dubious. In fact it is the largest problem this blockchain is facing [11].

6.2.3 Ethereum Transactions Trend Chart

The NEO blockchain (which can also handle intelligent deals) can handle 10,000 transactions every second. If developers of Ethereum fail to address this scalability issue, companies may think about using other blockchain networks to host their dApps and intelligent deals instead of the Ethereum blockchain network. If this happens in the future of Ethereum, its price will probably crash even further.

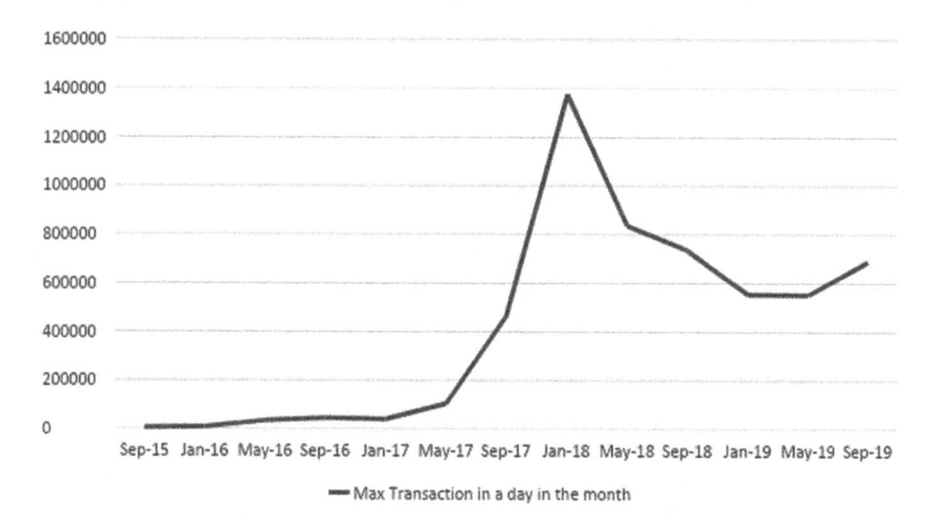

FIGURE 6.7 Ethereum Transactions Trend Chart [7].

Fortunately, developers of Ethereum are aware of these issues and they are making some important changes. Let us consider the possible solutions:

i. **Proof of Stake**

Ethereum blockchain uses a proof-of-work consensus approach. When working on a proof, miners must utilize extra computing power available to them to solve really complex puzzles. If you have a very powerful hardware device, you have the best chance of winning a mine bonus. Everyone tries to solve the puzzle at once, and finally the miner who solves it first will be the winner.

Proof of Stake is a very different approach because not all the miners try to solve one problem at a time. Instead, they operate one after the other. They are selected at random. However, you need to use a certain number of ethers to choose from.

The number of transactions they can reduce/check depends on the amount they decide. If a miner has a hundred ETH, he can deduct a hundred ETH worth of transactions in total. As per the scenario, this means that they can mine mines until the value of the transaction is reduced to the amount they have determined.

The proof-of-stake approach brings various benefits to the Ethereum blockchain network. The first is an increase in energy efficiency. A proof-of-work network requires a lot of energy because miners all work on the same puzzles and waste their computing power. On the other hand, proof-of-stake works with a single system, which means that overall, very little energy is consumed.

The future of Ethereum with a proof-of-stake approach also minimizes the risk of centralization. Moreover, proof-of-work allows different groups of miners to pool their available resources to improve their total mining reward odds. But the problem here is that a few people, especially those who control large pools of miners, have a lot of impact on the network. However, proof-of-use makes it very difficult to achieve.

The Proof of Work by Ethereum Developers is called the "Casper Project" led by Vlad Zamfir.

ii. **Plasma**

During August 2017, Vitalik Buterin announced the Plasma project for the very first time. This project is being developed initially to address the scalability issues which are being faced by Ethereum. Plasma is basically a protocol that actually eliminates the need for the Ethereum blockchain network to process unwanted data. This is done by creating a second level above the main block chain.

Smart contracts can still be processed. However, they will be published on the blockchain once the deal is over. This significantly reduces the amount of processing the blockchain needs to validate transactions and also saves much disk space.

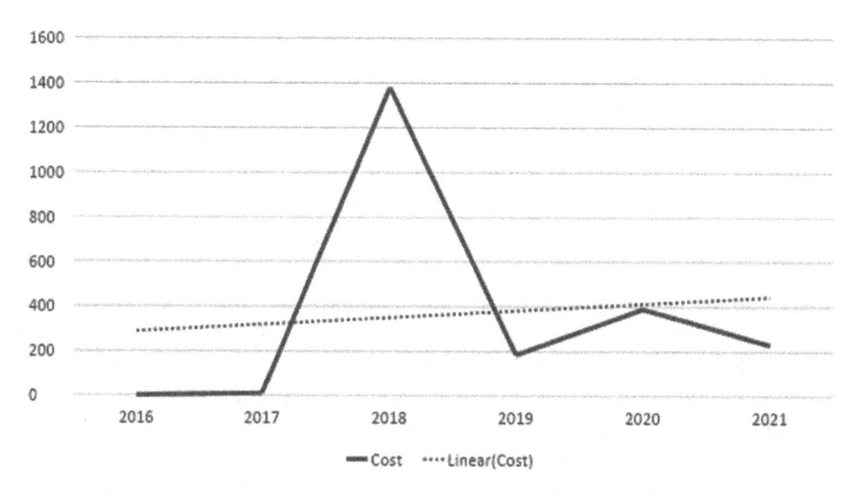

FIGURE 6.8 Ethereum Forecast, Long-Term Price Analysis [7].

The Plasma Protocol also speeds up transaction time, which in turn allows network dApps to be hosted on the network without slowing down the whole system.

The Plasma project is in early development stages, therefore information about when it will be installed on the Ethereum blockchain network is not available.

iii. Sharding

Sharding was also developed to solve scalability problems, similar to the plasma protocol. Before we understand what sharding is, let us try to understand the actual problem. Currently, every node which is connected to the Ethereum blockchain network must check each and every transaction that goes through the network. For example, if you need to review a hundred transactions in the next block, each node should check every hundred transactions.

While this technique is good for security measures, the speed of the network is the same as that of the connected individual nodes. The parts do things differently. After installation, the network is divided into various smaller parts and each part is known as shard. Each and every shard has distinct transaction histories in which each node operates on its own. Since each node is not required to verify each transaction, it is expected that it will significantly increase the number of transactions which can be handled by Ethereum blockchain.

We know that Ethereum can cost more if one of the above solutions is successfully implemented. This is because although only 15 transactions can be handled by the Ethereum network per second, it is still easily the second most important blockchain network in the industry. If you can increase that number to thousands of transactions per second, consider how much this blockchain network will work.

As we mentioned early on in this chapter, there are already blockchains that can handle thousands of transactions per second, and others are being created. But can Ethereum do this? Of course! On the other hand, cryptocurrency markets are still at an early stage, so there is no guarantee of what will happen. Talking about future regulation is a problem. For example, Japan controlled the cryptocurrency industry as much as it controlled its financial services industry. As a result, Japan is one of the largest cryptocurrency trading volumes and is accepted by more than 200,000 different businesses.

Apart from this issue per unit price, Ethereum also has some other issues. One of the biggest problems is government regulations that are not well defined and under preparation for the blockchain network. This makes these rules very confusing and insecure. For Ethereum, regulatory uncertainty eased somewhat on June 14, 2018, when William Hinman, the SEC's corporate finance director, said Ethereum was not currently safe. "If the token or coin-operated network is sufficiently decentralized," the assets are not necessarily investment contracts, because the buyers do not reasonably expect someone to perform important administrative functions [14].

Despite facing so many hurdles, Ethereum is continuously developing. This blockchain technology has the world's largest active community among all blockchain technologies. As per our discussion it is almost clear that currently the hurdle of regulations has been crossed and with the development of protocols like Plasma, Sharding etc it seems that blockchain will certainly overcome this issue as well. It is abundantly clear that Ethereum will develop but the future being the future, it is uncertain, but the prospects are looking excellent.

6.2.4 Solidity and Other Technologies

Ethereum transactions are based on smart contracts which are like normal contracts and are automatically binding. There are several languages in which smart contracts can be coded. One of the famous languages is Solidity. Ethereum transactions are executed on Ethereum Virtual Machine (EVM) just like Java and JVM. Each machine running EVM is referred to as a node in the Ethereum blockchain.

The working of any Ethereum-based application is as follows:

a. First a smart contract is written using Solidity etc.
b. Then it is converted into Bytecode which is further converted to processor-level opcodes by EVM.
c. Then this contract is sent to the Ethereum network as a transaction.
d. Subsequently this transaction is mined by all nodes. Now this smart contract is deployed on the network and a public address is allocated to it.
e. In order to interact with this contract, a transaction is sent to its address, specifying function which has to be invoked.
f. This method call is then saved in blockchain after the transaction is mined. A cost is associated with every method call. This cost is computed as 'Gas' and its price is expressed in 'Ether' and paid by the sender of the transaction.

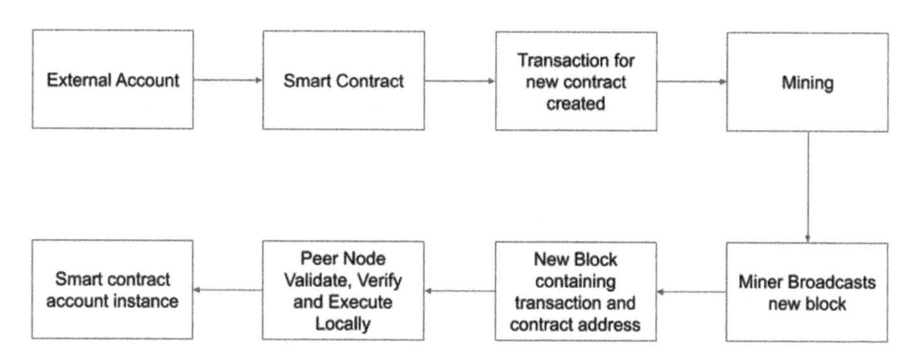

FIGURE 6.9 Ethereum Transaction Flow.

Now in this whole process several technologies are involved; let us discuss these technologies:

i) **Solidity**
This is an object-oriented and high-level programming language which is used for writing intelligent contracts. Solidity can also be used to execute intelligent deals on different blockchain platforms, especially Ethereum. It was developed by a number of former collaborators for Ethereum to write intelligent contracts on blockchain platforms such as Christian Reitweisner, Gavin Wood, Alex Bereggazzi, Yoichi Hirai, Liana Husickan and the Ethereum foundation. Currently, the primary language in Ethereum for writing smart contracts is Solidity, as well as in many other private and open-source blockchain networks running on platforms which compete with the Ethereum blockchain network. For example Hyperledger Burrow blockchain and Monax. A researcher at Cornell University explained that the infamous 2016 DAO hack was not able to happen due to direct vulnerability or a flaw of the DAO contract but due to security holes in solidarity contracts. During that incident EVM worked as intended. Those security holes were not due to bad practices by developers; they existed in the fundamental design of Solidity. Solidity is a programming language for developing EVM hosted smart contracts, which is consistently typed. As Wood noted, it relies on ECMAScript syntax to alert existing web developers. ECMAScript does not have static typing and different return types but Solidity has. In contrast to other languages such as Mutan, Snake etc which target EVM, there are very significant differences in Solidity. Contracts are supported by complex member variables, including arbitrary hierarchical associations and structures [13].

ii) **Ethereum Virtual Machine**
The implementation of powerful virtual sandbox stack and contract byte code embedded in each complete Ethereum node is made possible by Ethereum Virtual Machine, EVM. Smart Contracts are generally and frequently written in high-level languages such as Solidity and then this code or contract is compiled by EVM to EVM bytecode.

This ultimately implies that the network, file system or other processes of the host system are completely separate from the machine code and bytecode. Each node in the Ethereum blockchain network runs an Ethereum Virtual Machine instance which allows it to accept those same machine code and instructions. Ethereum Virtual Machine is complete in itself. This means that EVM represents a complete system that can perform every logical step of the computation function [13][15].

iii) Smart Contracts

Smart contracts are basically a set of computer code between two or more systems or individuals which are executed on the blockchain networks, with a pre-agreed set of terms by the parties involved. If these predefined rules are complied with at the time of execution, the smart contract can also be executed to generate output. Smart contracts validate, verify and enforce the terms of contracts thus making distributed automation of contracts possible. We can transparently exchange valuables such as money, stocks, property, etc. without the need for an intermediary and the system is free of conflict by using smart contracts.

In simpler words, we can say that these are automatically executable codes with fixed rules and are stored in the blockchain network. The code executes automatically and produces output if a predefined set of conditions is fulfilled. Smart contracts are very useful in various business cooperations. Smart contracts are used to accept the terms and conditions established by consent of both sides in any business deal. Ultimately this reduces cost and risk involved

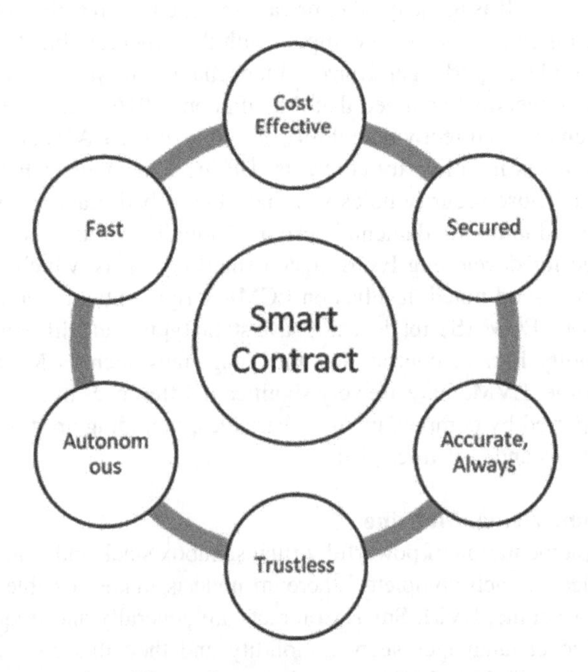

FIGURE 6.10 Properties of Smart Contracts.

in various deals as automated enforcement diminishes the risk of fraud and the absence of third-party involvement reduces costs [12].

In summary, smart contracts are usually based on the way digital assets are shared with multiple parties, and participants can manage their assets automatically. These assets will be credited to the participants in accordance with the terms of the agreement and will be redistributed. Smart contracts can track performance in real time and save costs. Smart contracts are therefore fast, cost-effective, secure, autonomous and accurate.

6.3 APPLICATIONS OF BLOCKCHAIN

Blockchain is a new technology and hence it has a lot of undiscovered areas. The impact rate of blockchain technologies on the existing marketplace is around 25%. Due to the adaptability and innovativeness of blockchain applications it has become the buzzword of the modern world. The most widely used practical implementation is in cryptocurrencies and digital identity management. Various other applications are discussed below.

6.3.1 Blockchain in Online Marketing

Advertising has changed a great deal in the previous decade, yet it is going to experience another development, thanks in enormous part to blockchain. Truly, while a large portion of us partner advanced advertising with things like AI and examination, blockchain might be the most troublesome innovation yet to hit advertisers in each industry. Blockchain is changing advanced advertising, and you might be astonished who will profit.

As the computerized world is moving toward AI and AI advertising has hit a significant storm, all credit goes to blockchain. While a large portion of us were tackling our math problems utilizing the basic Pythagoras hypothesis, blockchain has changed and disturbed the progressing advanced promoting organizations.

The principal purpose for such an achievement of blockchain is that it empowers start-to-finish exchanges between two gatherings without the need for any outsider as a controller. This is accompanied by reduction of additional charges, and thus organizations which use this technology benefit.

While different innovations help legislatures and private organizations, blockchain has given the client authority and a playing field for the intensity of getting verified information back. For instance, the Brave program is another sort of program that has changed the manner in which clients cooperate with adverts. Rather than just filling the entire screen with entries, clients opt into survey promotions and get Basic Attention Tokens (BATs) for the advertisements with which they cooperate which basically builds the profitability of the advert organizations and the client can also freely decide to quit. This sort of thinking has changed the understanding of clients and has given them the perfect amount of room for their information which helps both the client and the organization itself. We need these sort of creative thoughts that invades somebody's close-to-home space yet takes care of the issue [6].

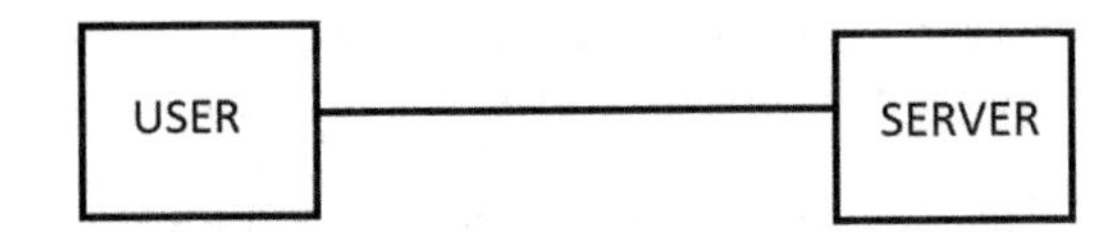

FIGURE 6.11 End-to-end user server interaction.

Another blockchain-based innovation, Blockstack, is made to ensure the advanced privileges of the clients by making another sort of decentralized organization for clients which does not divulge their information to different information-hungry organizations.

Finally it is true that if you are in marketing industry you might not love this concept at first but as it grows in the upcoming years and becomes the next go-to thing it can lead to much more profitable businesses and happy customers. So using it might be the best possible aspect for growing companies.

So we should say hello to transparency, and goodbye to spying companies.

6.3.2 Blockchain and Machine Learning

Well these days everybody thinks about AI; even a multi-year old youngster can give to you a couple of examples. This interesting issue has changed product firms altogether, and this can likewise be utilized joined at the hip with blockchain innovations. It can expand security for encryption, it can test for blunders during exchanges and significantly more tasks that require a large number of calculations and time.

The models of AI can be applied to any application containing an immense measure of information and time span in which the model can be prepared to give ideal outcomes to the client.

Let us take our current brought together data management system, explicitly the ones that gather and store a great deal of information, for example, Google or Instagram; these organizations store their information in significant servers and one way or another a small amount of information becomes mixed up in the stack heap. Information being the most important thing these organizations take major aggressive stands over one another. Subsequently utilizing a decentralized system may disturb this challenge since AI models will upgrade and tackle the information putting away issue and blockchain will give the client all the authority over their information [6].

AI + BLOCKCHAIN = BETTER MODELS

Let us envision a spot with decentralized frameworks administered by AI models. How simple it would be for the client to enter and associate with the framework and get the ideal outcome at a moment.

A genuine guide to examine the impacts of AI on blockchain systems can be spam location. Let us state that you have more than a large number of messages in your decentralized post box and not many of them are utilized by programmers for phishing or different purposes. For this situation, the utilization of a blockchain is huge as we trade a great deal of data about the occasions and subsequently it will thus expand the precision and learning capacities of the AI model for foreseeing a bothersome occasion by utilizing past information given to it. Thus one can endure a computerized assault from a programmer utilizing AI and thus the profitability increments.

This prompts the conclusion that AI has a great deal of favorable circumstances in the event that we use it with blockchain frameworks.

One can use it to improve

- **Storage –** Teaching the machine the best possible way to store data will result in faster transactions and hence blockchain systems will get much more speed.
- **Security –** Giving the machine learning models the task to find the encryption algorithms and handling the security of massive transactions will lead to a major increase in security of systems and greatly improve blockchains.

6.3.3 Blockchain and Decentralized Web Network

Digital forms of money furnish individuals over the globe with instant, secure, and frictionless cash, and blockchains give a lasting record stockpiling to their exchanges. Earlier frameworks expected clients to confide in a focal position that the fiscal stockpile and installment move will not be tampered with. Blockchain advancements out of date this strategy for installment move by giving a trustless domain so that there is never again a need to depend on an outsider to guarantee your installment moves, along these lines making a person-to-person (peer-to-peer) condition.

The very meaning of blockchain says that it is a decentralized record that can store data safely and permanently, using cryptographic encryption and hashing. In any case, it appears actually, the word 'decentralized' is one way or another restricted uniquely to the definition. A great many blockchains out there in the market utilize unified systems [5].

Nobody controls blockchains and they do not have the infrastructural main issue of disappointment. Henceforth, they are politically and structurally decentralized. Be that as it may, they are coherently incorporated since they carry on like a solitary PC.

Be that as it may, regardless of whether we pass by the above definition, are blockchains as they are today decentralized?

Anyway, that implies unified blockchains are bad?

Not really, and this is on the grounds that blockchains fill different needs, and these may expect them to be concentrated.

As indicated by the Crytpoasset Taxonomy Report, just 16% of digital forms of money are completely decentralized. The different digital forms of money looked into are either brought together, or just semi-decentralized. Just 9% of every utility token was seen as adequately decentralized and just 7% of money related resources; for example, those conceived from beginning coin contributions are decentralized. Digital forms of money, for example, Bitcoin, Litecoin, Stellar that act fundamentally as a methods for installment are among the most decentralized sorts of crypto resources, as per the report.

While the first digital money—bitcoin—was intended to be decentralized and removing the control of governments, a few specialists guarantee that even bitcoin cannot be named as completely decentralized since most of the bitcoin miners are from China.

It is regularly said that blockchain can be considered as another 'layer of trust', that is being included 'on top' of the web. This is valid from multiple points of view. Much the same as the web takes into account immediate and prompt trade of data and information, the blockchain takes into consideration immediate and quick trade of significant worth without a concentrated and believed outsider or middlemen. The term 'trust' can be comprehended in the broadest sense: cash, property, licenses, commitment, notoriety, time, work, and so forth. Though the web required brought together and believed foundations so as to guarantee the respectability and authenticity of exchanges and the exchange of significant worth, blockchain systems will assume control over this job starting now and into the foreseeable future.

Over more than fifteen years, exchange went on the web and social communications in the broadest sense have moved into the cloud. Most organizations offer administrations or access to products through applications on the web. In any case, the business rationale of concentrated administrations on the web and of organic market of internet business has prompted a gathering of intensity and control by a set number of organizations. They acknowledged and took care of installments and account holders the executives, kept up a stage administration or gave access to administrations, substance and merchandise or sorted out coordinations and satisfaction.

It appears to be very clear and authentic that organizations need control; not only to play out their business exercises, but also for lawful reasons. With offering and running applications, interfacing individuals or keeping up budgetary or different business exchanges, come legitimate duties and commitments. Most organizations need to know their clients and have the ability to bar people from their administrations or decline to offer to them in any case. Clients, then again, trust known and set up brands with their cash, information and security. Much of the time they value their administrations. In different cases they rely upon them, similar to when gaining admittance to a ledger.

Thinking about the upsides and drawbacks of brought-together administrations on the web, their duties and power uncover a somewhat sensitive and delicate foundation. It is no occurrence that blockchain and dispersed record advances show up on the business scene when the defects of the current brought-together set-up are dwarfing its advantages. Proof for this can be found by taking a look at the financial part, online business or web-based life stages: trust is a value that does not stick, yet must be earned and supported in a repeating way. What is more, this is the place blockchain becomes possibly the most important factor.

Blockchain can possibly disturb the intensity of brought-together organizations, organizations and stages simultaneously. As decentralized business develops, new stages will emerge that will work totally extraordinary in various manners. Rather than taking care of and encouraging exchanges through their concentrated records, they will bolster shared exchanges with different digital forms of money or tokens on various blockchains. Rather than depending on the power and notoriety of known and trusted undertakings and its brands, exchanges will be founded on 'trustless trust', which is built up on the blockchain. No outsiders will be expected to direct exchange any more, no banks will be required for clearing budgetary exchanges or the exchange of significant worth. Universal deals will be conceivable without national money transformations. Exchanges will be borderless, consentless and oversight safe. What is more, nonetheless, rather than depending on unified administrations to ensure access to substance, data or information, clients will in the long run be responsible for their records and become genuine proprietors of their benefits.

6.3.4 Why Blockchain is Taking over the Internet

Another year has started. Thinking back to 2017 it was really a wild ride on the crypto rollercoaster. Alt coins experienced huge development and consideration. Some portion of this can be credited to bitcoin's wild ride ever highs, trailed by huge remedies. Fortunately, ideally, this is behind us.

Maybe bitcoin was the symbol of atonement. Maybe, the ideal specimen. What bitcoin most certainly did, was acquaint digital currency and the blockchain with the world. Regardless of whether individuals got it or not, they threw cash at it. Right up till the present time trades still have shortlists for new clients, institutional cash is gradually crawling into the market, and the deluge of capital does not seem, by all accounts, to be easing back at any point in the near future.

Since the start of 2017 the business has seen some crazy development.

Bitcoin advertise top $18 bil → $736 bil [USD]

Complete advertise top (barring bitcoin) $2 bil → $481 bil [USD]

Without precedent for history this gives the whole digital currency / blockchain industry a market top well over $1 trillion dollars!

Something intriguing that you may have seen from the qualities above is that the remainder of the business is developing (as far as market top) impressively quicker than bitcoin. Actually, in 2017 the remainder of the business became six times quicker than bitcoin.

To jump somewhat deeper, we can take a look at the level of complete market top of bitcoin contrasted with the remainder of the business. Since December 1, 2017 bitcoin's share of the overall industry has diminished from 55% to around 33% which has given path for many other blockchain advantages for break into the space.

If I were a betting man, in 2018 I would hope to see Ethereum surpass bitcoin concerning share of the overall industry. Without question, Ethereum will keep on beating bitcoin to the extent the group and innovation goes. Bitcoin's administration issues, hesitance to develop, and amazingly high expenses may simply lead it to its downfall.

This being stated, I do not think bitcoin is going to bite the dust totally. Bitcoin will never kick the bucket. What is more, for the situation that it does, at that point at any rate it will be prevailing by its hard-forked posterity, Bitcoin Cash. Which, I would make reference to, has an undeniable possibility and surpassing Old Man Bitcoin.

Blockchain innovation self-discipline Web 3.0, which I will allude to as 'the new internet'.

A computerized cash is an extraordinary confirmation of idea for blockchain innovation, yet without a doubt, it is one of the most fascinating. Blockchain innovation is genuinely very astonishing, and every day we are seeing new, imaginative organizations shaping. As it at present stands, the extent that I am concerned, the market is broken into three segments.

1. Cryptographic money (Bitcoin, Litecoin, Ripple)
2. Stage (Ethereum, NEO, IOTA)
3. Utility Token (TRON, EOS, Status)

Up to this point, the spotlight has been fundamentally on digital currencies; however, I feel that in 2018 the stage and utility tokens will turn into the core interest. Both of these parts, all the more explicitly the stage, are vital on the grounds that they help characterize the up and coming age of web. This up and coming age of web, or Web 3.0., will be contained a stack a lot of like the following:

1. A decentralized exchange layer. (Ethereum — being the most grounded)
2. A decentralized document stockpiling layer (IPFS and Swarm)
3. A decentralized informing layer (Matrix or Whisper)
4. A high throughput registering asset (Golem)

As should be obvious, the principal part of the riddle is the exchange layer, or as referenced over, the stage. The main blockchain stage will be the essential structure hinder for the new web, and as the present fight goes, Ethereum is winning.

With the entirety of this stated, it will be an extremely intriguing year. Decentralized applications will keep on flooding into the market, and we may see the stage for the web of things to come harden itself inside the market. Get your popcorn, it will energize.

6.4 CONCLUSION

In this chapter, we began with a brief introduction of blockchain. We went through the basic concepts which form the base of this technology. Then we went through Ethereum, a platform for implementing Decentralized Applications (Dapps). Further, we discussed the applications of blockchain and how it is gradually taking over all the major fields.

Blockchain is basically a distributed ledger which works on top of a peer-to-peer network. It can be used to track anything of value in such a way that information is difficult to tamper with. As the name suggests, blockchain is a chain of blocks where every block has a cryptographic hash of the previous block, a timestamp and transaction data. The concept of hashing and consensus algorithm makes data resistant to modification. Then we discussed a comparison of transaction without blockchain and transaction involving blockchain, with the help of an example. Then we had a brief introduction of the key concepts responsible for the success of this technology such as decentralization, consensus, immutability, security and reliability. We saw how the idea of blockchain emerged in 1991 when Stuart Haber and W. Scott

Stornetta tried to create a system to prevent tampering of time-stamped documents. However, blockchain finally came into existence in 2008 due to Satoshi Nakamoto's white paper "Bitcoin: A Peer-to-Peer Electronic Cash System". For implementing blockchain, we need to be familiar with its key attributes. Some of the prerequisites for blockchain development include knowledge of one or more object-oriented programming languages and basic understanding of topics like cryptography and data structures. We then discussed consensus algorithms and how they provide a mechanism to reach an agreement among nodes in the blockchain network. Now since blockchain is difficult to implement from scratch and even more difficult to test due to the requirement of blockchain accounts and currencies, we can use tools and platforms which ease the process of blockchain development.

After discussing blockchain in detail we discussed Ethereum, its future and technologies associated with it. Ethereum was launched in 2015 and is a blockchain open-source software platform that uses its own cryptocurrency ether. It enables the creation and implementation of Smart Contracts and Distributed Applications ("Dapps") without time consuming, fraud, scrutiny or third-party intrusion. Ethereum is not just a platform, but a programming language (Turing Complete) that works on the blockchain, helping developers to create and publish distributed applications. In Ethereum transactions are based on smart contracts which are like normal contracts but are also autonomous in nature. These transactions are executed in Ethereum Virtual Machine or EVM just like Java-JVM. Each machine in the network or internet on which EVM instance is active and listening to method calls is known as node. Every transaction request is mined on all nodes. This virtually makes the Ethereum network a distributed supercomputer. The smart contracts which are used in Ethereum are generally written in solidity language. Smart contracts can be created using Golang, Lisk, etc. A Smart contract is essentially a protocol to digitally facilitate, verify and enforce the negotiation or performance of a contract.

REFERENCES

1. An overview of blockchain technologies: Principles, opportunities and challenges Gültekin Berahan Mermer; Engin Zeydan; Suayb Sb Arslan 2018 26th Signal Processing and Communications Applications Conference (SIU)
2. Advanced Applications of Blockchain Technology - Editors Shiho Kim, Ganesh Chandra Deka
3. Blockchain – Wikipedia - https://en.wikipedia.org/wiki/Blockchain
4. What are the key concepts of blockchain development – www.leewayhertz.com/blockchain-development-key-concepts/
5. An Overview of the Emerging Technology: Blockchain Rishav Chatterjee; Rajdeep Chatterjee 2017 3rd International Conference on Computational Intelligence and Networks (CINE)
6. Blockchain Beyond Bitcoin: Blockchain Technology Challenges and Real-World Applications Muniba Memon; Syed Shahbaz Hussain; Umair Ahmed Bajwa; Asad Ikhlas 2018 International Conference on Computing, Electronics & Communications Engineering (iCCECE)
7. Ethereum Official Docs: ethereum.org
8. A Beginner's Guide to Ethereum Feb 23, 2017 - Linda Xie

9. What is Ethereum? Ameer Rosic; blockgeeks.com
10. Blockchains: How they Work and Why they'll Change the World Sept 28, 2017 – Morgan Peck
11. Ethereum: A secure decentralised ledger Dr Gavin Wood; Founder Ethereum and Ethcore
12. A Survey of Attacks on Ethereum Smart Contracts Nicola Atzei, Massimo Bartoletti, and Tiziana Cimoli Universit à degli Studi di Cagliari, Cagliari, Italy
13. Introducing Ethereum and Solidity - Foundations of Cryptocurrency and Blockchain Programming for Beginners, Authors: Chris, Dannen
14. Understanding Ethereum, Bitcoin's Virtual Cousin; Nathaniel Popper NYT
15. The Ethereum Virtual Machine — How does it work? Luit Hollander; Developer bij MyCrypto

7 Internet of Things (IoT) Applications with Blockchain Technique

Ram Akuli

CONTENTS

7.1 INTRODUCTION TO THE INTERNET OF THINGS (IOT)

7.2 WHAT IS IOT?

The Internet of Things, or IoT, is a system of interrelated computing devices, mechanical and digital machines, objects, animals or people that are provided with unique identifiers (UIDs) and the ability to transfer data over a network without requiring human-to-human or human-to-computer interaction.

FIGURE 7.1 Example of IoT.

IoT is creating opportunities for more direct integration between the physical world and computer-based systems, and resulting in improved efficiency, accuracy and economic benefit.

7.3 BENEFITS OF IOT

The Internet of Things offers a number of benefits to organizations, enabling them to:

- Monitor their overall business processes.
- Improve the customer experience.
- Save time and money.
- Enhance employee productivity.
- Integrate and adapt business models.
- Make better business decisions.
- Generate more revenue.

7.4 EXAMPLE OF IOT

Examples of objects that can fall into the scope of Internet of Things include connected security systems, thermostats, cars, electronic appliances, lights in household and commercial environments, alarm clocks, speaker systems, vending machines and more as shown below in Figure 7.1.

7.5 APPLICATIONS OF INTERNET OF THINGS (IOT)

1) *Smart Home*

Wouldn't you love it if you could switch on air conditioning before getting home or switch off lights even after you have left home? Or unlock the doors to friends for temporary access even when you are not at home. Do not be surprised with IoT taking shape companies are building products to make your life simpler and more convenient.

The Smart Home has become the revolutionary ladder of success in the residential spaces and it is predicted smart homes will become as common as smart phones.

The cost of owning a house is the biggest expense in a homeowner's life. Smart home products promise to save time, energy and money. Smart home companies like Nest, Eco bee, Ring and August, to name a few, will become household brands and are planning to deliver a never seen before experience.

2) *Wearables*

Wearables have experienced an explosive demand in markets all over the world. Companies like Google and Samsung have invested heavily in building such devices. But how do they work?

Wearable devices are installed with sensors and software which collect data and information about users. This data is later pre-processed to extract essential insights about the user.

These devices broadly cover fitness, health and entertainment requirements. The pre-requisite from Internet of Things technology for wearable applications is to be highly energy efficient or ultra-low power and small-sized.

3) *Connected Cars*

The automotive digital technology has focused on optimizing vehicles' internal functions. But now this attention is growing toward enhancing the in-car experience.

A connected car is a vehicle which is able to optimize its own operation and maintenance as well as the comfort of passengers by using onboard sensors and internet connectivity.

Most large auto makers as well as some brave startups are working on connected car solutions. Major brands like Tesla, BMW, Apple, and Google are working on bringing the next revolution in automobiles.

4) *Industrial Internet*

The Industrial Internet is the new buzz in the industrial sector, also termed as Industrial Internet of Things (IIoT). It is empowering industrial engineering with sensors, software and big data analytics to create smart machines.

According to Jeff Immelt, CEO, GE Electric, IIoT is a "beautiful, desirable and investable" asset. The driving philosophy behind IIoT is that smart machines are more accurate and consistent than humans in communicating via data. And this data can help companies pick up inefficiencies and problems sooner.

IIoT holds great potential for quality control and sustainability. Applications for tracking goods, real time information exchange about inventory among suppliers and retailers and automated delivery will increase supply chain efficiency. According to GE the improvement industry productivity will generate $10 trillion to $15 trillion in GDP worldwide over the next fifteen years.

5) *Smart Cities*

The Smart City is another powerful application of IoT generating curiosity among the world's population. Smart surveillance, automated transportation, smarter energy management systems, water distribution, urban security and environmental monitoring all are examples of Internet of Things applications for smart cities.

IoT will solve major problems faced by people living in cities such as pollution, traffic congestion and shortage of energy supplies etc. Products such as cellular communication-enabled Big Belly trash products will send alerts to municipal services when a bin needs to be emptied. By installing sensors and using web applications, citizens can find available parking slots across the city. Also, sensors can detect meter-tampering issues, general malfunctions, and any installation issues in the electricity system.

6) *IoT in Agriculture*

With the continuous increase in the world's population, demand for food supply has greatly increased. Governments are helping farmers to use advanced techniques and research to increase food production. Smart farming is one of the fastest growing fields in IoT.

Farmers are using meaningful insights from the data to yield better return on investment. Sensing for soil moisture and nutrients, controlling water usage for plant growth and determining custom fertilizer are some simple uses of IoT.

7) *Smart Retail*

The potential of IoT in the retail sector is enormous. IoT provides an opportunity to retailers to connect with the customers to enhance the in-store experience.

Smartphones will be the way for retailers to remain connected with their consumers even out of store. Interacting through smartphones and using Beacon technology can help retailers serve their consumers better. They can also track the consumer's path through a store and improve store layout and place premium products in high-traffic areas.

8) *Energy Engagement*

The power grids of the future will not only be smart enough but also highly reliable. The smart grid concept is becoming very popular all over the world.

The basic idea behind smart grids is to collect data in an automated fashion and analyze the behavior of electricity consumers and suppliers for improving efficiency as well as the economics of electricity use.

Smart grids will also be able to detect sources of power outages more quickly and at individual household levels like nearby solar panel, making possible a distributed energy system.

9) *IoT in Health Care*
Connected healthcare still remains the sleeping giant of the Internet of Things applications. The concept of connected healthcare system and smart medical devices has enormous potential not just for companies, but also for the well-being of people in general.
Research shows IoT in healthcare will be massive in coming years. IoT in healthcare is aimed at empowering people to live a healthier life by wearing connected devices.
The collected data will help in the personalized analysis of an individual's health and provide tailormade strategies to combat illness.

10) *Poultry and Farming*
Livestock monitoring is about animal husbandry and cost saving. Using IoT applications to gather data about the health and wellbeing of cattle, ranchers knowing early about the sick animal can extract it and help prevent a large number of sick cattle.

7.6 IOT AS A NETWORK OF NETWORKS

Figure 7.2 below shows these networks connected with added security, analytics, and management capabilities. This will allow IoT to become even more powerful in what it can help people achieve.

7.7 THE FUTURE OF IOT

As far as the reach of the Internet of Things, there are more than 12 billion devices that can currently connect to the internet, and researchers at IDC estimate that by 2020 there will be 26 times more connected things than people.

7.8 INTRODUCTION OF BLOCKCHAIN

7.8.1 What is Blockchain?

A blockchain is a record of transactions. The name comes from its structure, in which individual records, called blocks, are linked together in a single list, called a chain. Blockchains are used for recording transactions made with cryptocurrencies, and have many other applications.

Each transaction added to a blockchain is validated by multiple computers. These systems, which are configured to monitor specific types of blockchain transactions, form a network. They work together to ensure that each transaction is valid before it is added to the blockchain. This decentralization of computers ensures a single system cannot add invalid blocks to the chain.

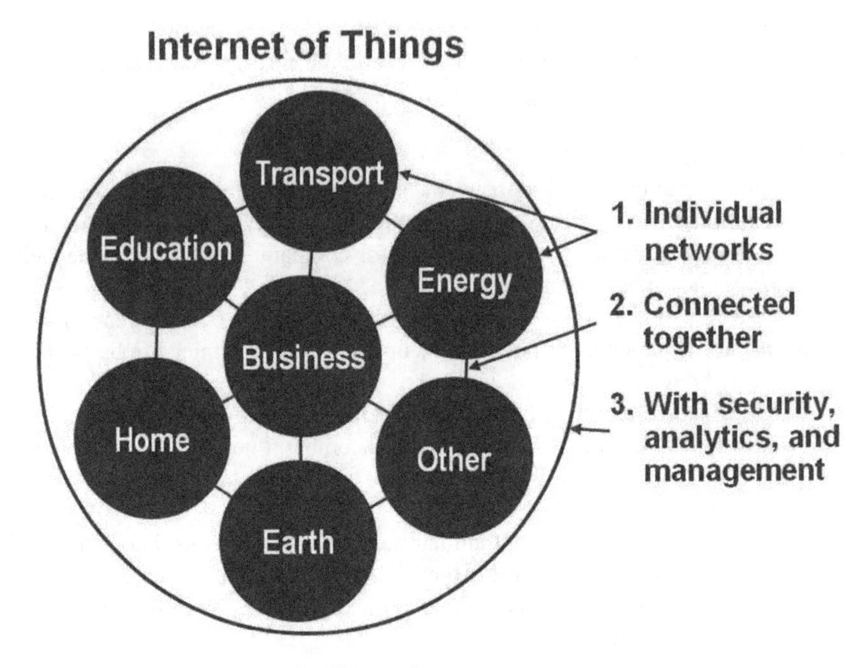

FIGURE 7.2 IoT as a Network of Networks.

FIGURE 7.3 Future of IoT.

When a new block is added to a blockchain, it is linked to the previous block using a hash generated from the contents of the previous block. This ensures the chain is never broken and that each block is permanently recorded. It is also intentionally difficult to alter past transactions in the blockchain since all subsequent blocks must be altered first.

7.8.2 Architecture of Blockchain

- Let us study the blockchain architecture by understanding its various components:

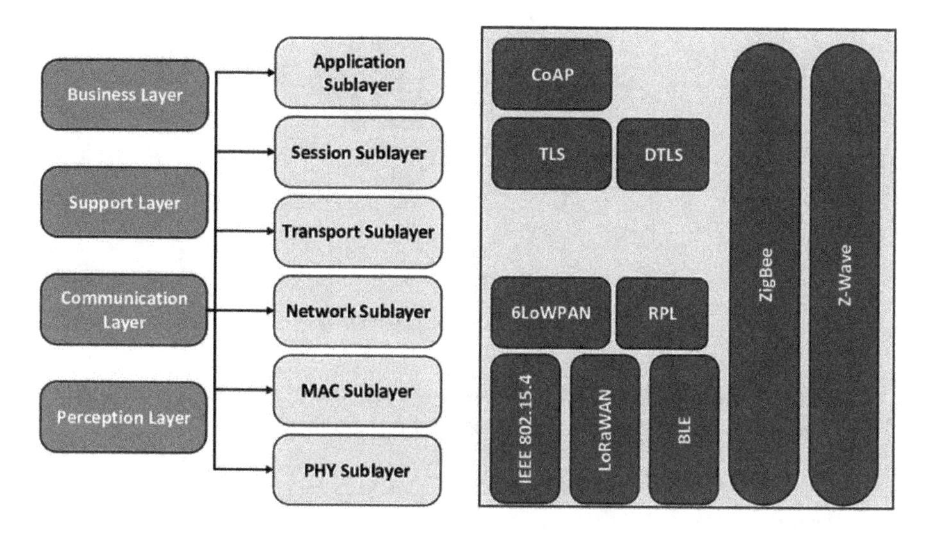

7.8.3 Genesis Block

- A blockchain is a chain of blocks which contain information. The data which is stored inside a block depends on the type of blockchain.
- For example, A bitcoin block contains information about the sender, receiver, and number of bitcoins to be transferred.

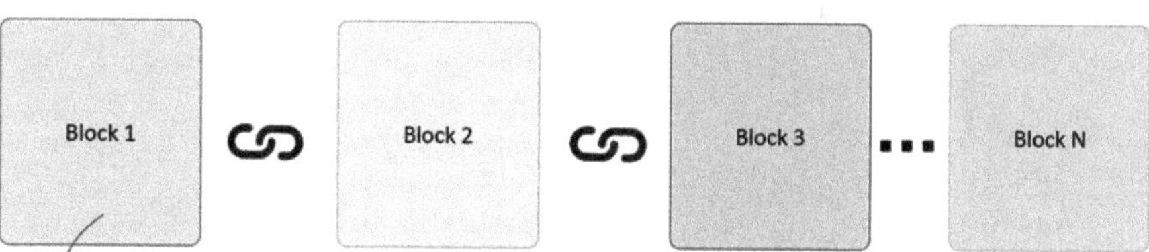

- The first block in the chain is called the Genesis block. Each new block in the chain is linked to the previous block.

7.8.3.1 Understanding SHA256 Hash

A block also has a hash. This can be understood as a fingerprint which is unique to each block. It identifies a block and all of its contents, and it is always unique, just like a fingerprint. So once a block is created, any change inside the block will cause the hash to change.

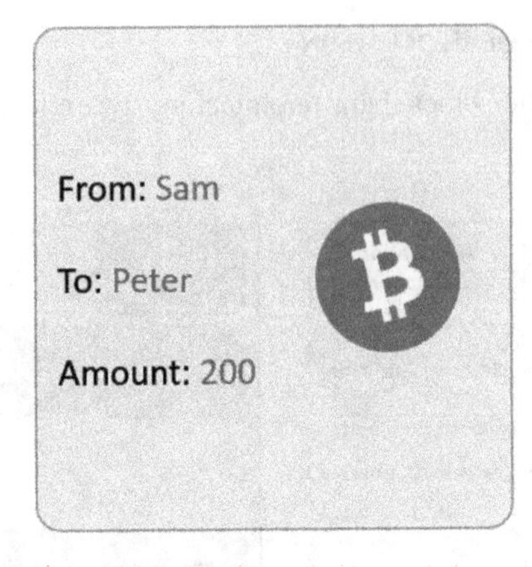

Bitcoin Block Example

Therefore, the hash is very useful when you want to detect changes to intersections. If the fingerprint of a block changes, it does not remain the same block.

Each block has

1. Data
2. Hash
3. Hash of the previous block

Consider the following example, where we have a chain of three blocks. The first block has no predecessor. Hence, it does not contain a hash of the previous block. Block 2 contains a hash of block 1, while block 3 contains a hash of block 2.

Hash acts as a Unique Fingerprint of the Block

Hence all blocks are containing hashes of previous blocks. This is the technique that makes a blockchain so secure. Let us see how it works:

Assume an attacker is able to change the data present in Block 2. Correspondingly, the hash of the block also changes. However, Block 3 still contains the old hash of

Block 2. This makes Block 3, and all succeeding blocks, invalid as they do not have the correct hash of the previous block.

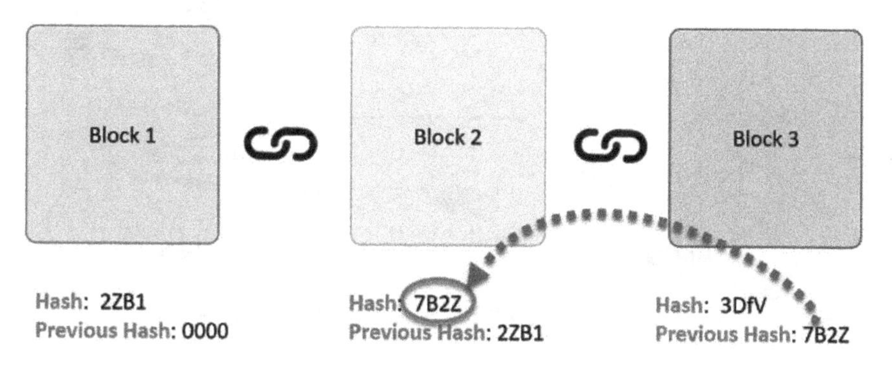

Therefore changing a single block can quickly make all the following blocks invalid.

7.9 APPLICATION OF BLOCKCHAIN

Blockchain is used for the secure transfer of items like money, property, contracts, etc. without requiring a third-party intermediary such as a bank or government. Once data is recorded inside a blockchain, it is very difficult to change it.

7.9.1 Wireless 2.0—Integrated Networks on the Blockchain

Wireless mesh networks as shown in the figure below, an emerging technology, may bring the dream of a seamlessly connected world into reality.

You would be forgiven for thinking that wireless mesh networking is just another marketing bullet point for new Wi-Fi routers, a phrase coined to drive up prices without delivering benefits.

However, we can avoid being cynical for once: mesh technology does deliver a significant benefit over the regular old Wi-Fi routers we have bought in years past and that remain on the market.

Mesh networks are resilient, self-configuring, and efficient. You do not need to interfere with them after often minimal work required setting them up, and they provide arguably the best and highest throughput you can achieve in your home. These advantages have led to several startups and existing companies introducing mesh systems contending for the home and small business Wi-Fi networking dollar.

7.9.2 Mesh Networks Solve a Particular Problem

Covering a relatively large area, more than about 1,000 square feet on a single floor, or a multi-floor dwelling or office, especially where there is no ethernet already present to allow easier wired connections of non-mesh Wi-Fi routers and wireless access points.

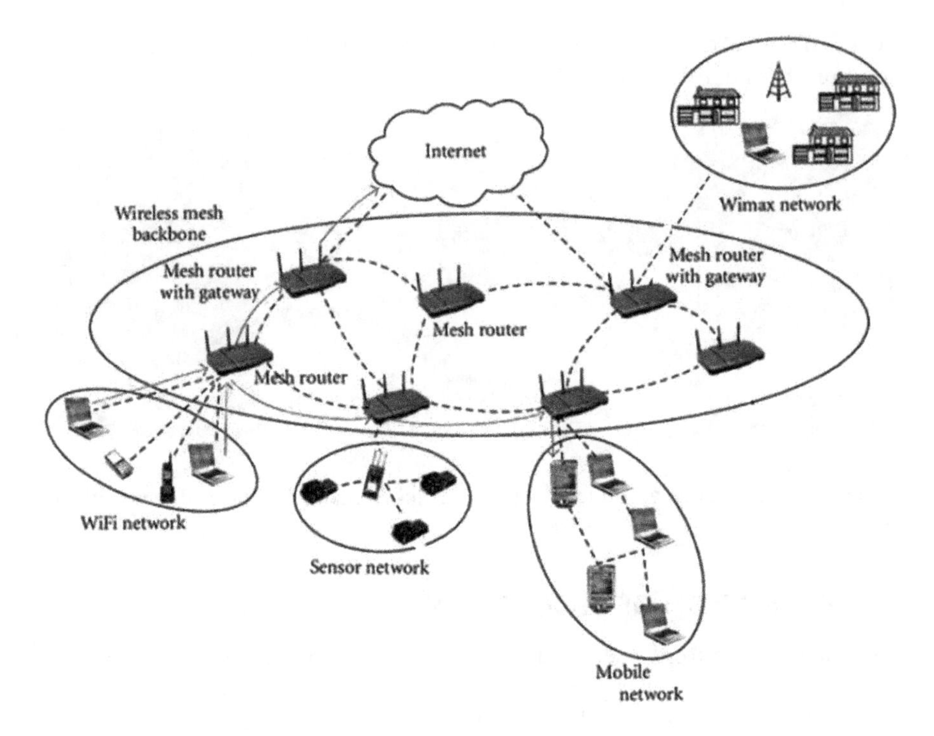

FIGURE 7.4 Wireless Mesh Network.

All the current mesh ecosystems also offer simplicity.

You might pull out great tufts of hair working with the web-based administration control panels on even the most popular conventional Wi-Fi routers.

In outdoor wireless networking, wireless mesh networks are the third topology after point-to-point and point-to-multipoint in order to build a wireless network infrastructure. Each device in a wireless mesh network is typically called a mesh node and is connected with multiple other mesh nodes at the same time.

Wireless mesh networks are also multi-hop networks because each mesh node can reach another node going through multiple hops and leveraging other nodes as repeaters. The major advantage of a wireless mesh network is its intrinsic redundancy and, consequently, reliability because a mesh network is able to reroute traffic through multiple paths to cope with link failures, interference, power failures or network device failures.

Two types of wireless mesh network are usually implemented for commercial and government applications:

- Unstructured or omni-directional wireless mesh networks
- Structured wireless mesh networks

In an unstructured wireless mesh network, each mesh node typically uses an omnidirectional antenna and is able to communicate with all the other mesh nodes that

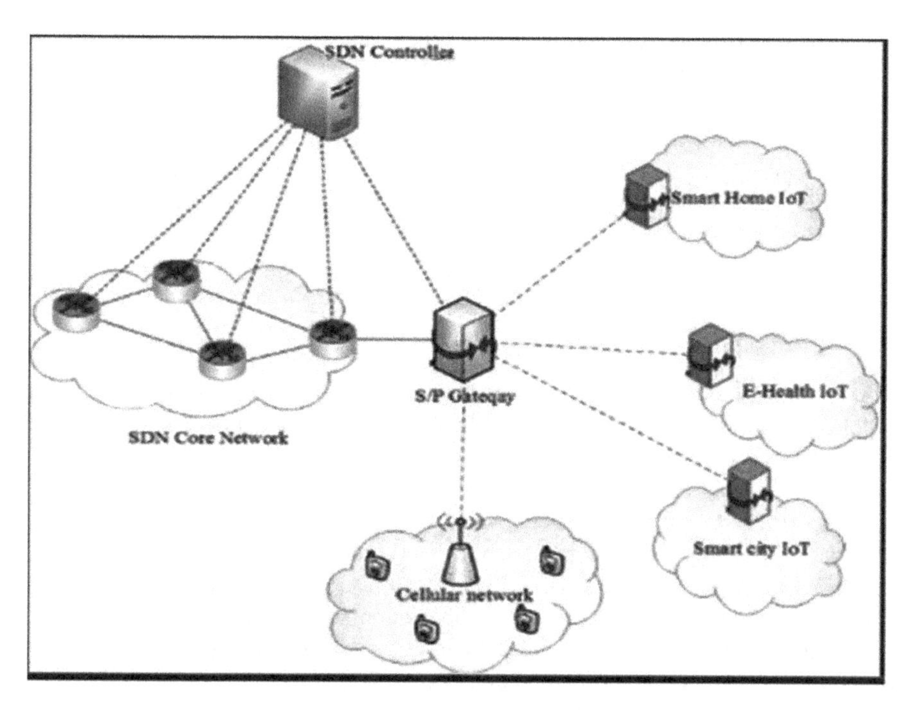

FIGURE 7.5 Point-to-Point, Point-to-Multipoint & Mesh Network.

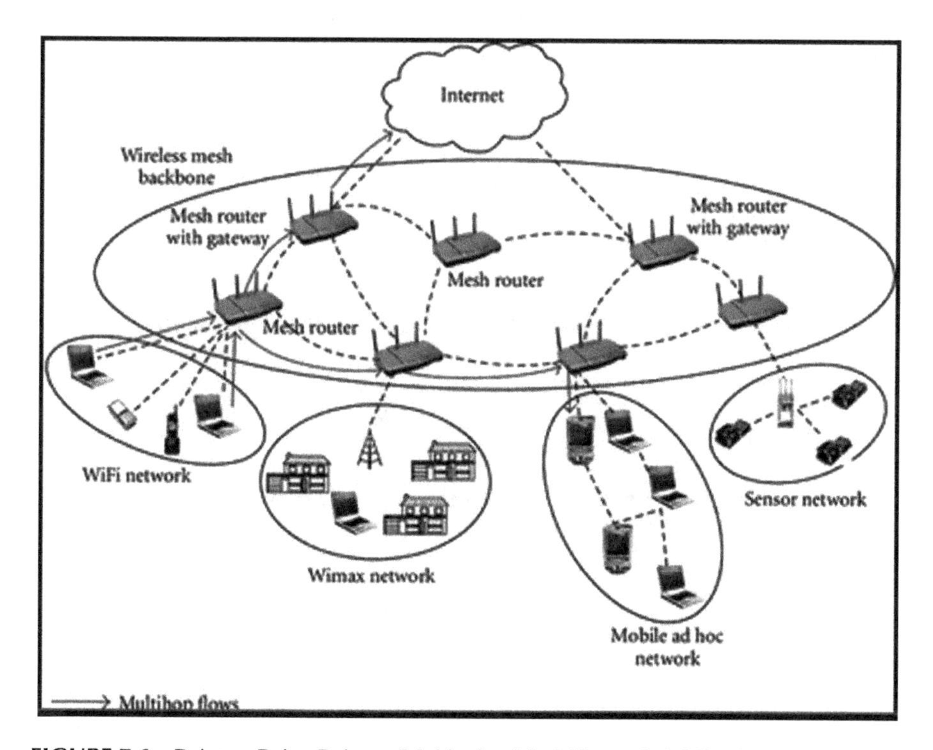

FIGURE 7.6 Point-to-Point, Point-to-Multipoint, Mesh Network & Mixed Network.

are within the transmission range. Wireless links in an unstructured wireless mesh network are not planned and link availability is not always guaranteed.

Depending on the density of the mesh network, there may be many different links available to other mesh nodes or none at all. Unstructured mesh networks are usually implemented with non-line-of-sight radios (NLOS) using low frequency and low bandwidth radios operating, for example, in the UHF bands, such as 400 MHz or in the license-free band at 900 MHz.

Unstructured wireless mesh networks leverage one single channel shared by all the radios. Therefore, the higher the number of hops a transmission requires, the lower the overall throughput of the network will be.

Structured wireless mesh networks are planned networks typically implemented using multiple radios at each node location and multiple directional antennas. A ring topology using multiple directional wireless links is commonly used in a structured wireless mesh network to enable each radio to seamlessly reroute traffic through different paths in the event of node or link failures.

Structured wireless mesh networks are often used for mission-critical applications such as wireless video surveillance, public safety, and industrial automation.

They provide the ideal network architecture in case a site requires a highly reliable and available wireless network for a broadband application such as video, voice and data streaming. Each link in a structured wireless mesh network operates on an independent channel and, therefore, the number of hops for a specific transmission does not affect the overall throughput of the network.

Wireless mesh networks have been studied in academia since the early 90s, initially mainly with military applications in mind, and then they started to get significant commercial traction between 2005 and 2010.

7.10 BITCOIN

A type of digital currency in which a record of transactions is maintained and new units of currency are generated by the computational solution of mathematical problems, and which operates independently of a central bank.

7.11 DIFFERENCE BETWEEN BITCOIN AND BLOCKCHAIN

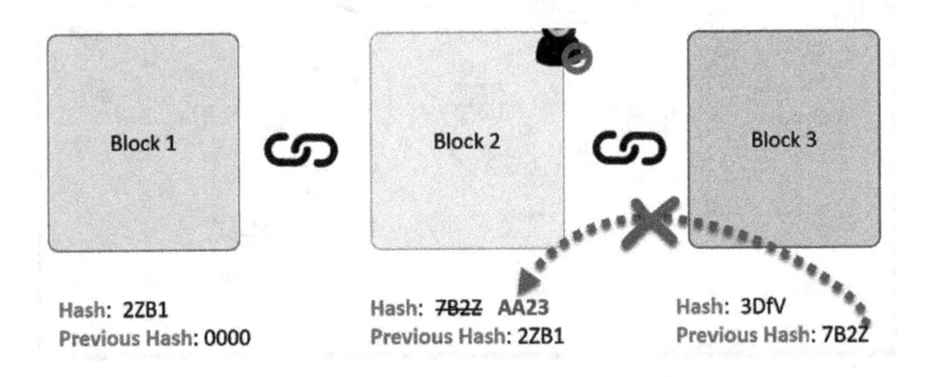

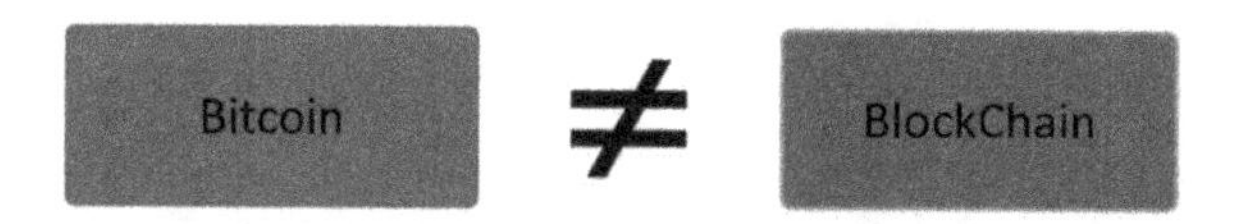

- Blockchain is not bitcoin, but it is the technology behind bitcoin.
- Bitcoin is the digital token and blockchain is the ledger to keep track of who owns the digital tokens.
- You cannot have bitcoin without blockchain, but you can have blockchain without bitcoin.

8 Security and Privacy-Enhancing Technologies for Blockchain and Cryptocurrency

Debasis Gountia and Utkalika Satapathy
CET (BPUT) Bhubaneshwar, India

CONTENTS

8.1 INTRODUCTION

Among recently technological advances, blockchain technology is an emerging new approach in the domain of information technologies. Blockchain is the use of advanced cryptographic proficiencies to implement a distributed system by a decentralized ledger of all existing transactions across a peer-to-peer (P2P) network, and allow fast processing of transactions in potentially trustless surroundings. It has firmly caught the imagination of cryptonerds, researchers, central bankers, and programmers, as well as politicians. By design, blockchain is a distributed decentralized tamper-proof ledger of records. Using blockchain technology, parties can control transactions without the need for a central certifying authority. Its potential applications allow fund transfers, voting, settling trades, and many other attractive uses. In blockchain, a transaction is the set of hash of the previous digitally signed transaction and the public key of the next owner. Each transaction is signed with a private key, and is verified by the public key [1], as shown in Figure 8.1.

Its working principle:

1. Anyone can request a transaction.
2. The requested transaction should broadcast to a P2P connection consisting of computers called nodes.
3. The nodes validate the transaction and the user's status using existing known algorithms.

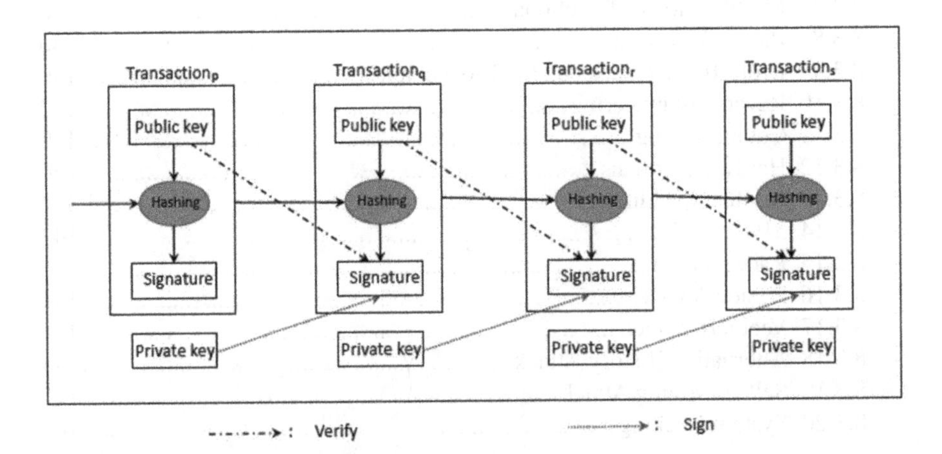

FIGURE 8.1 Network of transactions in a blockchain.

4. A validated transaction can exhibit cryptocurrency, records, contracts, and/or other important information. Cryptocurrency is a medium of exchange, generated and stored electronically in blockchain, using more secured encryption techniques to handle the generation of monetary units and to confirm the truth of the transfer of funds; bitcoin is the most suitable example.
5. Once validated, this transaction is merged with other transactions to generate a new block of information for the ledger.
6. The new data block is then combined to the existing blockchain permanently that is unalterable.
7. The transaction comes to a finish or an end.

In blockchain technology, data is stored in the form of multiple required blocks and these blocks are connected with each other through a network. A newly generated block would be connected to its former block; in this way this method creates a chain of blocks, which is called a blockchain. The process of adding new blocks to the blockchain is called mining [2]. The data stored in the block is permanent as it cannot be easily and directly changed. It is a very critical task to make any alteration or modification to the stored data. This is so because it needs agreement from all participating nodes for any update in the blockchain.

Each block of the blockchain consists of a hash of the previous block. A hash is the sequence of multiple characters and numbers. The features of transparency and verifiability prevent unauthorized access to the blocks and hence do not allow any changes. "No brainer" use cases are offered for applying blockchain technology by capital and finance markets. Bitcoin has proved itself successful in producing digital money and tracking its ownership. Today, there exist hundreds of cryptocurrencies. These blockchain technologies have become very attractive and popular due to the following facts of multi-activities in terms of privacy and confidentiality in the field of transactions:

- Support for all digital transactions
- Transparency
- Accurate tracking
- Cost reduction
- Provenance
- Permanent ledger: Creates an open permanent ledger, which makes it safe and easier to share information within the network
- Auditability
- Elimination of middleman: Avoids the need of a middleman which is able to reduce cost
- Faster time to market

Projects involving blockchain concepts should strive to prepare protocols in a manner such that their participants are incentivized to maximize the value of the system as a whole; in other words, it should be more profitable to secure and create the blockchain ecosystem more valuable than it is to cheat and make profit for oneself. This idea should be the essence for the design of the protocol underlying bitcoin's blockchain.

As the blockchain market has grown very quickly in the past few years, malicious people's attacks on the blockchain system are becoming a serious threat to transactions. Hence it is urgent to conduct research on the security issues of blockchain.

The remainder of this chapter is organized as follows. Section 8.2 presents related work on bitcoin scalability trade-off. Different effective attacks associated with blockchain are elaborated in Section 8.3 along with their potential defenses in Section 8.4. Comparisons and results analysis is presented in Section 8.5. Finally, conclusions are drawn in Section 8.6.

8.2 RELATED WORK ON BITCOIN SCALABILITY TRADE-OFF

Scalability is the strength of a system, process, or network to handle an increasing number of tasks with time, or its potential to be enhanced to adjust to that growth. For example, a network is considered scalable if it is capable of growing its total output when load is increased and resources such as hardware are merged with the system. Scalability is a substantial factor in computer systems, for example, databases, networking, and routers.

Bitcoin scalability trade-off refers to the discussion regarding the constraints on the number of transactions a bitcoin network can handle to process and execute successfully. It is related to the fact that records (known as blocks) in the bitcoin blockchain are limited in size and frequency. Blocks of bitcoin contain the transactions on the bitcoin network. The on chain transaction processing capacity of the bitcoin network is limited by the average block creation time of ten minutes and the block size limit. These jointly constrain the throughput of network. The transaction processing capacity maximum is estimated between 3.3 and 7 transactions per second. There are various proposed and activated solutions to address this issue efficiently.

Enhancing the transaction processing limit of a network demands various improvements to the technical principles of bitcoin, in a process known as a fork. Forks can be classified into two types: soft fork and hard fork.

8.2.1 SOFT FORK

A soft fork is any change of rules that enable recognition of newly produced blocks as valid by the old software. Thus it is backward-compatible. A soft fork is also able to split the blockchain when newly generated blocks not considered valid by the non-upgraded software and the new rules.

8.2.2 HARD FORK

In contrast to a soft fork, a hard fork is a software upgrade introducing new rules to the network, thus abolishing the old software that is not able to recognize new blocks as valid [3]. In case of a hard fork, all nodes meant to work in accordance with the new rules need to update their software.

If one group of nodes continues to follow the non-upgraded old software while the other group of nodes uses the new updated software, a split will take place. For example, platforms such as Ethereum, introduced in Vitalik Buterin's paper [4], that can allow for the production of smart contracts; digital entities with ingrained computer code that execute contractual agreements based on future events [5]. These entities represent financial instruments, currency, and land ownership, etc. Ethereum has hard-forked to make all the investors in the DAO, which had been hacked due to a vulnerability in its code [6]. In this case, the split creates Ethereum and Ethereum Classic chains by the fork.

In 2014, the NXT community considered a hard fork that could have led to a rollback of the blockchain records to mitigate the effects of a theft of 50 million NXT against a major cryptocurrency exchange. The hard fork proposal was rejected, and a few funds were retrieved after negotiations and ransom payment [7]. Alternatively, to assure from a permanent split, maximum nodes using the new upgraded software can return to the old rules, as was the case in the bitcoin split [8]. Bitcoin Cash is a hard fork of bitcoin enhancing the maximum block size. Bitcoin XT, Bitcoin Classic and Bitcoin Unlimited all supported an enhancement to the maximum block size through a hard fork.

Lei *et al.* [5] suggested a technique for secure key management in an Intelligent Transportation System (ITS). In [9], Khan *et al.* proposed that the intrinsic features of blockchain technology can be exploited to address many privacy and security related problems of IoT systems. In [10], a decentralized system is suggested which combines Inter Planetary File System (IPFS), Ethereum blockchain, and Attribute Based Encryption (ABE) to assure fine-grained access control to the owners and the users of the stored data.

Finally, Guo *et al.* [11] approach the combined blockchain with an Attribute Based Signature (ABS) mechanism to prevent collusion attacks in multiple authority parties.

8.2.3 Efficiency Improvements

Transaction throughput is limited practically by a parameter known as block size limit. Various increases to this limit, and proposals to remove it completely, have been proposed over bitcoin history.

8.3 EFFECTIVE ATTACKS ON BLOCKCHAIN

Blockchain has successfully started up a brave new world to create, hold, and distribute digital values in the world of business. Some are too afraid of blockchain to consider it to be the next wave of technology revolution; others dismiss this concept as a passing craze for the underworld of "crypto-cyber criminals". Figure 8.2 and Figure 8.3 summarize the different types of the emerging blockchain threats, vulnerabilities, and attacks that are described in the following sections. Many of the following problems and solutions described in this article are anticipatory in nature; we pose these problems based on our best knowledge of how current transaction systems work and extrapolate from the existing security literature. However, the ideas put forth in this research article are not intended to replace existing works. Instead, these hardware-based countermeasures can be used to bolster system security or provide assurances of security that would otherwise be unachievable to the blockchain world.

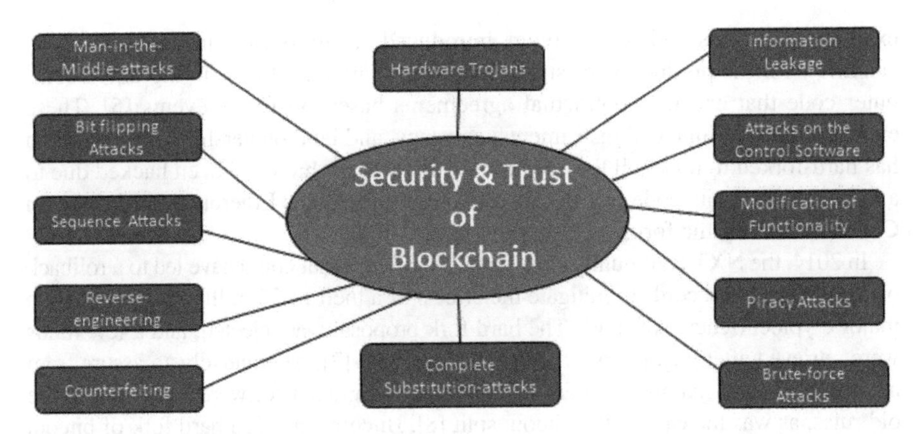

FIGURE 8.2 Blockchain security and trust consist of a diverse array of threats, vulnerabilities, and attacks.

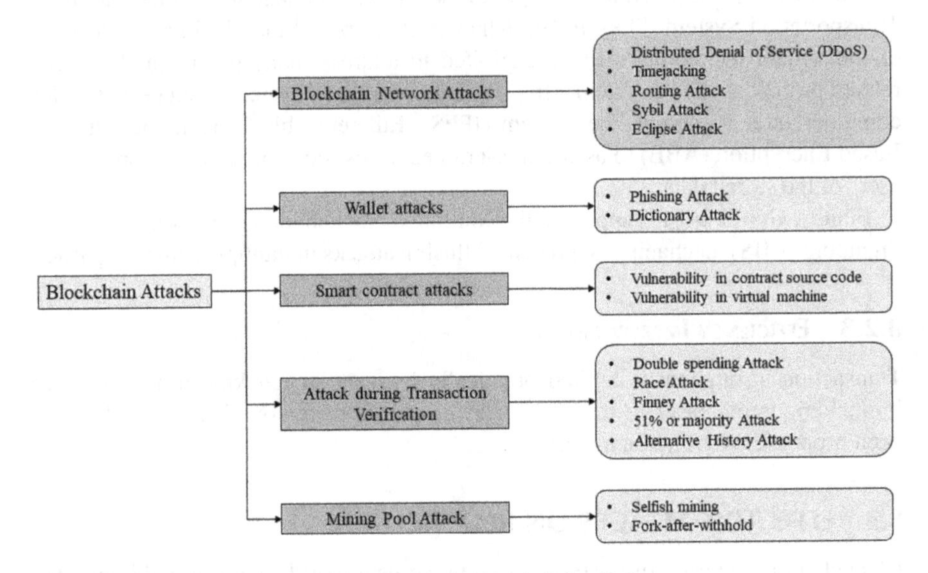

FIGURE 8.3 Different possibilities attacks on blockchain platform.

Blockchain technology is secured intrinsically. In blockchain, the data or ledger are distributed across several computers, and hence it has removed any single point failure. Furthermore, a blockchain is scarcely possible to hack due to the implementation of cryptographic proofs and consensus mechanisms such as game theory within it. With the above underlying security features; nevertheless, blockchain security issues still prevail.

8.3.1 Man-in-the-Middle Attacks

In the field of computer security, a man-in-the-middle attack is an attack where malicious people secretly relay and modify a transaction between two parties who believe

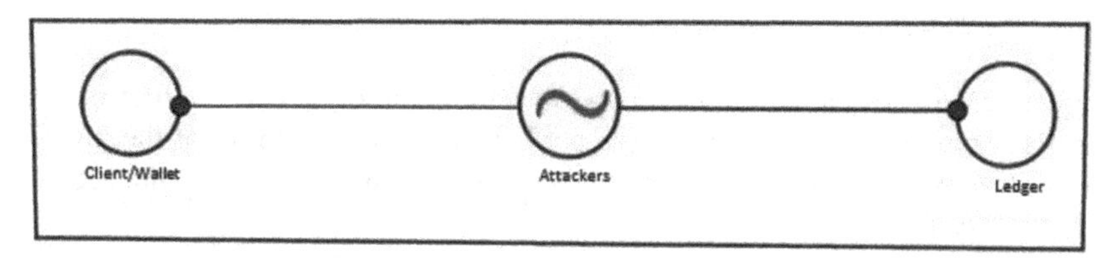

FIGURE 8.4 Man-in-the-Middle-attacks.

they are directly making a transaction with each other without any interference [12]. For example, active eavesdropping, where the attacker produces an independent connection between these victims and relays a transaction between them to produce trust that they are doing transactions directly with each other over a private network, when in fact the entire transaction is controlled by the malicious people as shown in Figure 8.4. These intruders intercept all relevant transactions passing between these victims and throw in either a new malicious one or alter the aforementioned transactions.

8.3.2 Bit-Flipping Attacks

Such types of attack are based on the substitution/replacement principle [13]. In a bit-flipping attack, a single bit of the transaction amount is modified which produces a significant error.

8.3.3 Sequence Attacks

Such attacks are also based on the substitution/replacement principle. In sequence attack, N-bits in the transaction can either be modified, inserted or deleted by an attacker. An intelligent adversary would be able to manipulate in such a way that most of the process proceeds normally.

8.3.4 Complete Substitution Attacks

Such attacks are also based on the substitution/replacement principle. A complete substitution attack is an attack in which the proposed transaction is completely replaced with an alternate one for a significant fault. This is the most extreme attack into transaction field.

8.3.5 Information Leakage

Attackers may disclose unauthorized the privileged information of different transaction involved in blockchain. Such examples of privileged information include client data, secret password, proprietary protocol, etc.

8.3.6 Attacks on Control Software

An unscrupulous coder can modify the error-recovery software in order to bypass the error-recovery mechanism. This is possible for both custom and general-purpose design flows of transactions.

8.3.7 Modification of Functionality

Attackers could maliciously force an unintended operation to execute. For instance, an attacker could subtly downgrade the performance and reliability of the functionality of blockchain, thereby depressing the end user's assurance and confidence in the blockchain system.

8.3.8 Piracy Attacks

There are protocols for different transaction applications. These are known as Intellectual Property (IP) of blockchains. Attackers can violate these IPs, for example, make a duplicate of the previous transactions and repeat the same again and again. Piracy of a transaction is an important unique factor of proprietary blockchains which security aspects require much efforts for which billions of dollars will be acquired. However, their piracy is not guaranteed to be well protected. Hence, traditional blockchains are vulnerable under IP thief threat as the attacker can easily pirate test protocols of transactions.

8.3.9 Brute-Force Attacks on the Blockchain

Attackers could try to their best to crack the confidential password of a transaction by applying all combinations of digits, letters, and special characters. This is known as a brute-force attack. Greater security guarantee is achieved by hardening resistance to brute-force attacks with high confusion and diffusion.

8.3.10 Reverse Engineering

Reverse engineering (RE) is the technique of analyzing a system to identify its components and their internal structures, interconnections, etc., and produce the representation of the system in another form or a higher level of abstraction [14]. RE is rigorously applied to disassemble a device in different ways such as cloning, duplicating, and reproduction. In this subsection, the RE of blockchain systems is acquired by extracting their internal physical structures and information using destructive techniques for secret information detection by foreign attackers.

8.3.11 Counterfeiting

A counterfeiting transaction is one which is a repeat of an already done transaction. Therefore counterfeit is a threat to blockchain like other attacks.

8.3.12 Hardware Trojans

A hardware Trojan (HT) is able to modify the circuit system of transaction or insert a malicious circuitry into the design to disable/destroy the whole system for a specific input/time. This HT is able to modify the designed circuit during either fabrication or design and cause unwanted behavior. These are also designed to disclose

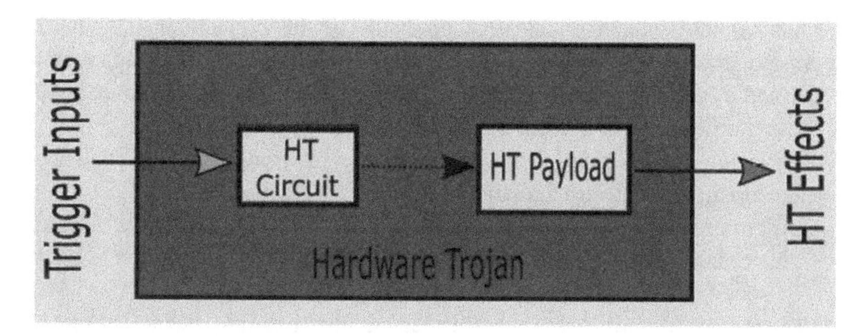

FIGURE 8.5 A typical structure of hardware Trojan.

the transaction secret information, Denial of Service, and alter system functionality. Attackers can insert HTs at any level from high-level system design specification to the transistor level of IC design flow [15].

A typical structure of HTs that could be inserted into a blockchain is shown in Figure 8.5. Some key terms related to HTs are with their meanings:

- Trigger: an event which initiates the HT. When this particular event starts, the HT circuit is automatic activated for deadly functionality.
- Payload: an event that activates the Trojan, responsible for implementing HT attacks, which could result in serious effects such as information leakage, denial of service (DoS), and blockchain reliability degradation.

Hardware Trojans can be inserted into blockchain as per the following categories:

- **Insertion phase**: Blockchain HTs can be inserted in any of the following phases:
- Specification: Blockchain HTs can maliciously alter the specification, for example, US Dollar (USD) to Euro (EUR) during runtime to make the transaction incorrect.
- Design: The blockchain designer can alter the transaction to alter the outcome.
- Fabrication: A hardware Trojan can be maliciously inserted during chip fabrication by tampering in the chip factory.
- Assembly: During the assembly of blocks, a malevolent integration engineer enacts the collection of blocks wrongly to produce erroneous output.
- Calibration and testing: A malicious tester can also insert an HT maliciously during the testing and calibration phase to overcome the blockchain concept.
- In-field: In the blockchain field, attackers falsify the transaction protocol by altering their agreement.
- **Abstraction level:** Hardware Trojans are able to insert at the following phases of abstraction level:
- System level: At system level, elements of individual domain and their interconnections are mentioned by the system engineer. Blockchain can be modified to result in erroneous output.

- Physical level: Each physical components of the blockchain, i.e., hardware components and wiring, chip platform and their locations and dimensions are defined at physical level. Hardware Trojans can be inserted by altering any of the aforementioned physical components and/or their dimensions.
- **Activation mechanism:** This describes the internal and external triggering mechanism of hardware Trojans.
    - Internal trigger: This trigger is executed for a particular instance of time slot.
    - External trigger: This type of trigger is executed externally due to output of a specific transaction.
- **Effect:** The effect of HTs is to alter ledger functionality, disclose secret transaction information, degrade performance, cause DoS attacks, and so on.

8.3.13 Double Spending

One of the major issue of a cryptocurrency developer is the problem of double spending. As per the name, this refers to when a person spends a coin multiple times, which creates an inconsistency between the spending ledger and the amount of available cryptocurrency. This happens when a blockchain network is tampered with and cryptocurrencies are stolen. The malicious node would send a copy of the transaction to make it looks legitimate, or might delete the transaction entirely as shown in Figure 8.6.

In a bitcoin network, there is a probability that a buyer can make a copy of the digital currency and send it to multiple retailers while keeping the original one. The most typical technique of double spending is when a blockchain hacker sends multiple transactions to the network and reverses the transactions, appearing as if those transactions never occurred. (www.investopedia.com/ask/answers/061915/how-does-block-chain-prevent-doublespending-bitcoins.asp).

This double-spending problem could be avoided in the bitcoin network or other blockchain-based cryptocurrencies by implementing a Proof-of-Work (PoW)

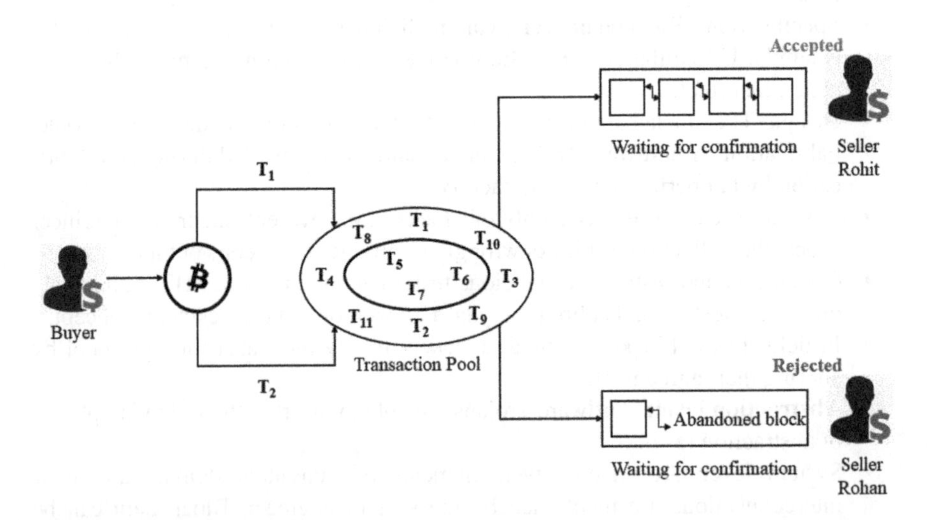

FIGURE 8.6 Double Spending.

consensus algorithm. This PoW is executed by miners who not only ensure the fidelity of past transactions on the blockchain's ledger but also detect and avoid double spending.

8.3.14 51% Attack

In a blockchain network, a 51% attack is a probable attack that can happen when an organization governs the mining power or so-called hash rate with a majority ratio. A bitcoin network is made secure by making all miners give consent on a shared ledger, i.e. blockchain. Every node in bitcoin ensures that these are working on a valid transaction at any point of time by looking at each other. Miners would have the potential to determine which transaction to give consent to given that the majority of miners are handled by a single entity. Hence this will give power to the miner to block other transactions and allow their own coins to be spent multiple times. This is also known as double spending as shown in Figure 8.7 [16, 17] (ref: https://learncryptography.com/ cryptocurrency/ 51- attack).

With this attack an attacker can perform the activities mentioned below:

- They can reverse his transactions that have already happened.
- They can block transactions from gaining any confirmations.
- They can block other miners to mine any other valid blocks.

However, an attacker cannot perform the activities below:

- They cannot reverse the transactions of others.
- Block transactions to be sent at all.
- Modify the number of coins generated per block.
- Generate coins out of nowhere.
- Send coins that were never owned by them.
- These attacks are valid till the attacker is in control (i.e., owns 51%). The transactions which had been turned down can be added just after the attacker loses their majority [18].

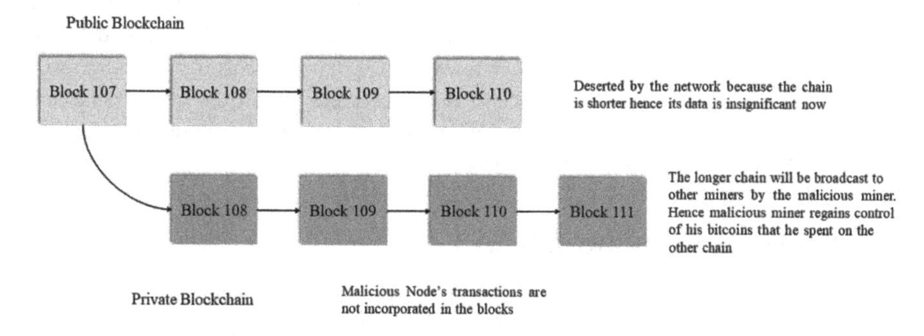

FIGURE 8.7 51% Attack Simplified.

If a blockchain network implements a Proof-of-Work (PoW) consensus mechanism, then it should have the proper security measures to avoid a 51% attack to be carried out. A few viable options are to be vigilant of mining pools, implementation of merged mining on a blockchain network with a higher hash rate, or utilizing a different consensus mechanism.

8.3.15 Race Attack

When a hacker sends two conflicting transactions in succession rapidly into the bitcoin network, this is known as race attack. This attack is comparatively easy to accomplish in blockchains which have utilized PoW as consensus algorithm. The dealer who receives payment instantly with "0/unconfirmed" status is vulnerable to reversal of transaction. An attacker performs a transaction by sending coins to the dealer directly, however, he will not wait for confirmation from the dealer and instantly sends a conflicting transaction (with the same coin used before for the dealer) to himself to the rest of the blockchain network. It is more likely that the conflicting transaction which happened later will be mined into a block and accepted by bitcoin nodes as valid node.

8.3.16 Finney Attack

An attacker first finds a dealer who accepts unconfirmed transactions. Then he performs a transaction to himself with the same amount to be sent to the dealer and find the block; however, he does not broadcast the block. Once the above step is completed the attacker will send the same amount to the dealer and wait till the item gets delivered. Finally, he broadcasts his previous block with the original transaction in it. This block will include the transaction that sent the coins to himself, hence the unconfirmed payment to the dealer will be invalidated. Moreover, the attacker will regain his coins and also the item for free.

8.3.17 Vector76 Attack

Vector76 attack otherwise known as one-confirmation attack. This attack is an amalgamation of the race attack and the Finney attack such that a transaction that even has one confirmation is still reversible. It is one of the variations of the double-spending attack. The attacker performs double spending by using the privately mined block during exchange. The wallet service such as exchange of cryptocurrency is vulnerable to this attack because of the acceptance of direct connections. If the Vector76 attack is successful then the attacker has to sacrifice one block because by not broadcasting it and only by broadcasting the attacked node.

8.3.18 Alternative History Attack

This attack occurs when an attacker initiates a transaction by sending coins to the dealer. Concurrently, the attacker will mine an alternative block privately which includes a conflicting double-spending transaction. The dealer will deliver the item to the attacker after waiting for m confirmations. Right away, if the attacker finds more than m blocks untie his chain and gets his coins back again; If not, he keeps trying to continue extending his chain of blocks with the hope of being able to catch up with the network. When the attacker is not ever able to extend his private blockchain compared to the public blockchain, then the attack fails.

8.3.19 Selfish Mining Attack

A selfish mining attack happens when an attacker (selfish miner in this case), does not broadcast a valid solution to the rest of the network. Rather than act like a regular miner and publish blocks to the network instantly after finding them, the attacker selectively releases blocks, or publishes many blocks all at once thus forcing the rest of the network to discard their blocks and lose revenue. The primary motives of selfish mining are to obtain an unfair reward which is bigger than their share of computer power spent, and confuse other honest miners and lead them to waste their resources in the wrong direction as shown in the Figure 8.8.

8.3.20 System Hacking

It is relatively hard to hack and tamper the records stored in a blockchain; however, the programming codes and systems that implement its technology can be vulnerable. The largest Tokyo-based bitcoin exchange, i.e., MtGox, was hacked in March 2014, and bitcoins worth $700 million were stolen. The reason behind this attack was ill-maintained and outdated codes which allowed hackers to perform double-spend. More recently, a DAO (Decentralized Autonomous Organization) that holds large quantities of Ethereum was exploited through a software vulnerability and the hacker stole $50 million worth of Ethereum [19].

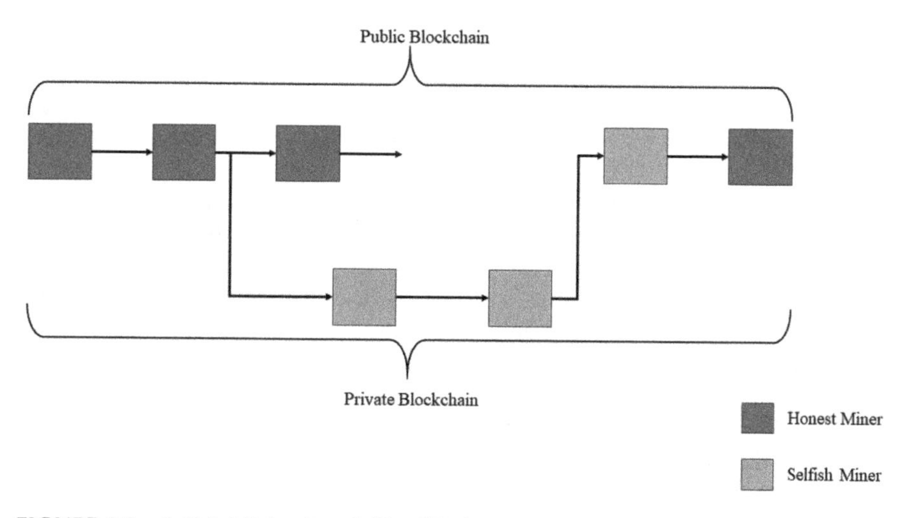

FIGURE 8.8 Selfish Mining Attack Simplified.

8.3.21 Illegal Activities

An illicit group of people can utilize the blockchain network or platform to perform various illegal activities. For instance, the Silk Road website was an online marketplace for illegal drug where sellers and buyers whose identities are anonymous did business using bitcoin (Hong 2015). Cryptocurrency that uses blockchain technology may also facilitate money laundering. Although bitcoin is not yet treated as a fiat currency, it makes it possible to create an "underground" channel for illegal movement of funds within its network.

8.3.22 Identity Theft

Although Blockchains preserve anonymity and privacy, the security of assets depends on safeguarding the private key, a form of digital identity. If one's private key is acquired or stolen, no third party can recover it. Consequently, all the assets this person owns in the blockchain will vanish, and it will be nearly impossible to identify the thief. The consequences may be more devastating than identity theft in the offline world, where third-party institutions (e.g., credit card companies) or central authorities safeguard transactions, control risks, detect suspicious activities, or help find culprits. Also, current cryptography standards are not completely uncrackable (Swan 2015). With the advent of quantum computing, it is not impossible for cryptographic keys to be cracked quickly, demolishing the foundation of blockchain technology (Crosby *et al.* 2016).

8.4 POTENTIAL DEFENSES AGAINST SECURITY THREATS ON BLOCKCHAIN

Potential defenses which assure security against man-in-the-middle-attacks, hardware Trojans, etc., include the following schemes:

- **Watermarking:** Someone requests a transaction directly known as client or via something called a wallet. The transaction directly request from the client will deliver either a commit or by a process at the server where the transaction initiated known as the master. In watermarking, the original client's digital signature is provided with the request. Watermarks are able to assure ownership as these are much more difficult to identify and modify. Unfortunately, the watermarking technique is not able to guarantee security against hardware Trojans.
- **Metering:** In this technique, both the public signature of wallet/master and the client's digital signature are added to the transaction request as processing constraints. This metering scheme is not also able to assure protection against hardware Trojans as the attacker is able to create and hide a malicious Trojan in the circuit due to availability of the design functionality.
- **Side-channel fingerprinting:** This scheme is able to detect hardware Trojans easily as the manufactured parametric characteristics, such as power, area, delay, and block characteristics of the transaction are compared with those of

statistical model. Any significant variation/deviation would be considered to be a Trojan. Side-channel fingerprinting is not able to assure authenticity or piracy, counterfeiting, or reverse-engineering attacks.

- **Reverse Engineering (RE):** This technique can also be constructively utilized to detect hardware Trojans. For RE, the state-of-the-art of blockchain should be made aware by the researcher to become successful in the detection of HTs inserted by foreign attackers. A typical RE flow should pass through de-packaging, delayering, and image processing of a blockchain.

  Mainly, its design and blocks are uncovered by RE scheme following the aforementioned steps is studied with a golden one (this means with no attack).

  This RE approach is both time-consuming and also destructive in nature. Hence, the RE technique is less applicable for HTs detection [20]. RE is generally used to assure about the Trojan-free blockchain used in the golden blockchain model development required for test time and runtime golden blockchain models.
- **Code analysis:** The code of blockchain functionality is analyzed to detect for any hardware Trojans inserted into the system. Also, any secured encryption algorithm and hash functions can be used for the confidentiality of transaction and hence protect from Trojan attacks on blocks. Code analysis is not able to protect the blockchain against piracy, reverse engineering, and counterfeiting attacks.
- **Obfuscation:** A code-obfuscation technique can be used by the blockchain designer for the mystification of transactions. This obfuscation is able to prevent hardware Trojan attacks indirectly as attackers would not be able to insert meaningful and stealthy hardware Trojans in such an obfuscated transaction sequences. Obfuscation is able to prevent hardware Trojans and reverse engineering, but not piracy or counterfeiting.
- **Locking:** A blockchain designer is able to add locks (i.e., digital multiplexers) which manage and control the flow of transactions among blocks or other blockchain components. These transactions will proceed further in a correct manner if and only if the correct secret key is applied, otherwise wrong

TABLE 8.1

Summary of potential defenses

Name of Defense	Name of Attack			
	Trojans	Piracy	Reverse Engineering	Counterfeiting
Watermarking	No	No*	Yes	No*
Metering	No	No*	Yes	No*
Side-channel Fingerprinting	No*	No	No	No
Reverse Engineering	Yes	No	---	No
Code Analysis	No*	No	No	No
Obfuscation	Yes	No	Yes	No
Locking	Yes*	Yes	Yes	Yes

transactions will proceed which results in an erroneous output. This key should be preserved in a tamper-proof memory in order to protect from vulnerabilities as the key is erased during reverse-engineering duration. Hardware Trojans are not able to be inserted since the blockchain functionality is hidden by the key. Locking prevents all aforementioned attacks: piracy, reverse engineering, and counterfeiting attacks, Trojans after fabrication, except for Trojans inserted during chip fabrication of the blockchain in industry.

8.5 COMPARISONS AND RESULTS ANALYSIS

In this section, all the potential defenses are summarized in Table 8.1 along with the statistics of comparisons among them. From the aforementioned Table, it is confirmed that the locking defense provides the best assurance for security issues in blockchain, followed by the obfuscation defense.

Depending on their business strategy and budget, companies and industry firms can choose any one or multiple aforementioned techniques to protect the blockchain against different known and existing attacks.

Symbols used in Table 8.1 meanings:

- Yes means both detection and prevention possible.
- Yes* means detect and prevent those Trojans inserted only after fabrication, but not those before fabrication.
- No means cannot detect, also not prevent.
- No* means only detection, but not prevention.

8.5.1 Discussion about Critical Infrastructures for Securing Blockchain

Because all the aforementioned techniques have their own pros and cons, one proposed direction is to use each for the highest HT coverage. For example, an RE-based scheme can guarantee a golden blockchain required for test time and run-time golden blockchain models. Side-channel and functional testing approaches are able to detect large and small HTs respectively that were inserted during chip fabrication. Runtime approaches can finally conclude to work as a last scheme of defense.

8.6 CONCLUSIONS

Though blockchain technology was designed to act as a backbone for crypto currency bitcoin from the beginning, blockchain is applied in other fields such as clinical diagnostic healthcare, government organizations, Intelligent Transportation System, etc., due to its open and decentralized framework, secure environment and tamper-proof characteristics. Though blockchain is a complex technology, it has had proved the potential to handle all record keeping processes, audit and assurance in the means transactions are initiated, processed, authenticated, recorded and reported at the time of demand with providing secured, trust and integrity. While blockchain technology cannot achieve its goal of other demanding features such as scalability, privacy and confidentiality. Hence it needs the attention of researchers as active areas of research and development due to the fact that these features are less matured. In the last few years, a number of cryptocurrencies, consensus protocols, and hashing functions have been developed in the networks. A few examples of the cryptocurrencies are NXT, Ripple, NEO, Cardano, Stellar, EOS, Litecoin, IOTA, Dash, Lisk, Zcash, Dogecoin, and many more. Finally, we hope that blockchain has the power to shape the 21st century.

Over the next decade, researchers will try a number of blockchain concepts and ideas, out of which some will succeed. But in the process some real-world problems will be solved and new businesses along with business models will emerge for the use of blockchain in better real-life state-of-the-art applications.

REFERENCES

1. D. Vujicic, D. Jagodic, and S. Randic, (2018) Blockchain Technology, Bitcoin, and Ethereum: A Brief Overview, March 2018, in 17th International Symposium INFOTEH-JAHORINA (INFOTEH), pp. 1–6.
2. F. Tschorsch and B. Scheuermann, (2016) Bitcoin and beyond: a technical survey on decentralized digital currencies, IEEE Communications Surveys & Tutorials, vol. 18, no. 3, pp. 2084–2123.
3. S. Nakamoto, (2008) Bitcoin: a peer-to-peer electronic cash system. Available at: https://bitcoin.org/bitcoin.pdf.
4. A. Castor, A short guide to Bitcoin forks, March 2017. Available at: www.coindesk.com/short-guide-bitcoin-forks-explained/.
5. T. Lee, (2013) Major glitch in Bitcoin network sparks sell-off; price temporarily falls 23%, Arstechnica.
6. V. Buterin, (2013) Ethereum white paper: a next generation smart contract & decentralized application platform, Available at: www.theblockchain.com/docs/Ethereum_white_paper_a_next_generation_smart_contract_and_decentralized application platform-vitalik-buterin.pdf
7. F. Coppola, (2016) A Painful Lesson For The Ethereum Community, Forbes.
8. C. M. Gillespie, (2016) Official NXT Decision: No Blockchain Rollback, Cryptocoin News.
9. A. Lei, H. Cruickshank, Y. Cao, P. Asuquo, C. P. A. Ogah, and Z. Sun, (2017) Blockchain-based dynamic key management for heterogeneous intelligent transportation systems, IEEE Internet of Things Journal, vol. 4, no. 6, pp. 1832–1843.
10. M. A. Khan and K. Salah, (2018) IoT security: Review, Blockchain solutions, and open challenges, Future Generation Computer Systems, vol. 82, pp. 395–411.
11. S. Wang, Y. Zhang, and Y. Zhang, (2018) A Blockchain-based framework for data sharing with fine-grained access control in decentralized storage systems, IEEE Access, vol. 6, pp. 38437–38450.
12. M. Conti, N. Dragoni, V. Lesyk, (2016) A survey of man in the middle attacks, IEEE Communications Surveys Tutorials 18 (3), 2027–2051.
13. J. Tang, M. Ibrahim, K. Chakrabarty, R. Karri, (2018) Secure Randomized Checkpointing for Digital Microfluidic Biochips, IEEE Transactions on Computer-Aided Design of Integrated Circuits and Systems, vol. 37, no. 6, pp. 1119–1132.
14. N. Jacob, D. Merli, J. Heyszl, and G. Sigl. (2014) Hardware Trojans: current challenges and approaches. IET Computers Digital Techniques, vol. 8, no. 6, pp. 264–273.

15. G. Maxwell, (2013) Coinjoin: Bitcoin privacy for the real world. Available at: https://bitcointalk.org/index.php?topic=279249.0.2013.
16. Baliga, Arati (2017) "Understanding blockchain consensus models." Persistent.
17. Watanabe, H., Fujimura, S., Nakadaira, A., Miyazaki, Y., Akutsu, A., & Kishigami, J. (2016, January). Blockchain contract: Securing a blockchain applied to smart contracts. In 2016 IEEE international conference on consumer electronics (ICCE) (pp. 467–468). IEEE.
18. Bastiaan, M. (2015, January). Preventing the 51%-attack: a stochastic analysis of two phase proof of work in bitcoin. Available at http://referaat. cs. utwente. nl/conference/22/paper/7473/preventingthe-51-attack-a-stochasticanalysis-oftwo-phase-proof-of-work-in-bitcoin. pdf.
19. Xu, Jennifer J. (2016) "Are blockchains immune to all malicious attacks?" Financial Innovation 2.1 (2016): 25.
20. S. E. Quadir, J. Chen, D. Forte, N. Asadizanjani, S. Shahbazmohamadi, L. Wang, J. Chandy, and M. Tehranipoor, (2016) A survey on chip to system reverse engineering, J. Emerg. Technol. Comput. Syst., vol. 13, pp. 6:1–6:34.
21. Bruice Schneier, Applied Cryptography, Wiley Press, Second Edition.
22. Douglas R. Stinson, Cryptography Theory and Practice, CRC Press, Second Edition.
23. Cryptocurrency Market Capitalizations, Available at: https://coinmarketcap.com/
24. A. Ali, M. M. Afzal, (2018) Confidentiality in Blockchain, International Journal of Engineering
25. Science Invention (IJESI), vol. 7, no. 1, pp. 50–52.
26. D. Shrier, W. Wu, A. Pentland, (2016) Blockchain & Infrastructure (Identity, Data Security), Available at: www.getsmarter.com/career-advice/wp-content/ uploads/2017/07/mit blockchain and infrastructure report.pdf.
27. C. Bao, D. Forte, and A. Srivastava, (2014) On application of one-class SVM to reverse engineering-based hardware trojan detection, in ISQED, IEEE, pp. 47–54.
28. R. Guo, H. Shi, Q. Zhao, and D. Zheng, (2018) Secure attribute-based signature scheme with multiple authorities for Blockchain in electronic health records systems, IEEE Access, vol. 776, no. 99, pp. 1–12.
29. Blockchain has the power to shape 21st century. Available at: https://economictimes.indiatimes.com/markets/stocks/news/blockchainhas-the-power-to-shape-21st-century/articleshow/ 65680293.cms
30. Mohamed Amine Ferrag, Makhlouf Derdour, Mithun Mukherjee, Abdelouahid Derhab, Leandros Maglaras, Helge Janicke, (2019) Blockchain Technologies for the Internet of Things: Research Issues and Challenges, IEEE Internet of Things Journal, in Press.
31. Leandros A. Maglaras, Ki-Hyung Kim, Helge Janicke, Mohamed Amine Ferrag, Stylianos Rallis, Pavlina Fragkou, Athanasios Maglaras, Tiago J. Cruz, (2018) Cyber Security of Critical Infrastructures, ICT Express (Elsevier), volume no. 4, issue no. 1, pp. 42–45.
32. Leandros Maglaras, Mohamed Amine Ferrag, Abdelouahid Derhab, Mithun Mukherjee, Helge Janicke, Stylianos Rallis, (2018) Threats, Protection and Attribution of Cyber Attacks on National Critical Infrastructures, EAI Transactions on Security and Safety, volume no. 5, issue no. 16, pp. 1–9.
33. D. Gountia (2019) Towards Scalability Trade-off and Security Issues in State-of-the-art Blockchain, EAI Endorsed Transactions on Emerging Topics in Security and Safety, vol. 5, issue no.18, pp. 1–9.

9 Security and Privacy in IoT

Neelamani Samal and Debasis Gountia
CET (BPUT) Bhubaneshwar, India

CONTENTS

The Internet of Things (IoT) plays a remarkable role in our modern daily life. With the use of IoT our daily life has become easier and dynamic. The IoT covers almost all fields, including home automation, office automation, healthcare, sports, industrial applications, and transport, Smart Cities, agricultural, Smart Energy and the Smart Grid, etc. Therefore this technology requires security management and privacy control over the user data which is in the cloud. Security in IoT is the technology area concerned with securing connected devices and networks in the Internet of Things environment where we have a number of devices connected over the internet with IP addresses. There are so many devices, sensors, things that we are wearing or using, things that we are interacting with that we will not even sense it in future. The IoT enables development of helpful tools and technological partners to resolve security issues which may occur in the future. Due to the small size of modern CPUs, the possible features of IoT and its applications are almost unlimited. IoT devices [16] are connected to the internet, so they are vulnerable to the same kinds of cyber-attack that can affect the user's personal and commercial information.

Through IoT the user is connected to a number of devices for gathering and controlling information. Therefore protecting consumer privacy of information has become difficult as the IoT becomes available to all involved. A number of devices are connected to different kinds of devices and this increase in connectivity and data collection results in less control of privacy of information to the public. Both the control of data and control of the different devices that are connected has become a major area of concern in the modern age of internet access. Therefore maintaining control over the data and device with maintaining privacy is of more concern as control can be lost if someone hacks into a smart phone or computer. Control can be lost when more and more companies collect data about users connected to the internet. Everything that we search, and all of our online activities, is being tracked by many companies that provide us with the services for their data-mining purposes and for improving user experience over the IoT. But sometimes these activities breach user security and privacy of data.

9.1 INTERNET OF THINGS (IOT) SECURITY TECHNOLOGIES

Creating a security framework [6] in IoT is purely problem specific i.e. depending on the case a specific security measure may be taken for improving security in IoT and to fix the shortcomings. Based on usage, it is possible to differentiate six main directions of security solutions for IoT.

In general among the greatest of the Internet of Things security issues, scientists suggest the following points are maximal in maintaining security. These issues include authentication problems, lack of proper data encryptions and security analytics, vulnerability of networks, and problems with the application program interface. The best part of security management is that these issues are successfully handled by the IoT experts to make it more user friendly and reliable, and each of them has a specific approach in terms of IoT security.

9.2 IOT SECURITY FOR NETWORKS

The best approach to creating a safe network with the collaboration of IoT is to connect it to a pre-installed backend system while maintaining security. To achieve the highest level of sophistication in the environment we combine traditional approaches of data protection such as antivirus and firewalls with protocols, standards, and complex device capabilities for specified uses.

9.3 AUTHENTICATION FIXING FOR IOT SECURITY ISSUES

Managing the authentication of accessing devices will definitely add a security benefit and can resolve most of the security threats. The functionality of these IoT security solutions ensures a safe authentication of one user into multiple networks or devices. Apart from the standard password-based procedure of maintaining security, the developers in IoT generally advise use of two-factor [15] authentications, biometrics, and digital certificates and other components that may be required to increase security.

9.4 IOT SECURITY TECHNOLOGIES FOR DATA ENCRYPTION

Data encryption can be considered one of the best ways to establish a secure environment or ecosystem for maintaining security. In this approach information becomes almost unreachable or unreadable for hackers with the help of standard and effective cryptographic algorithms designed and developed by the IoT expert. These tools are commonly used in a large range of applications. Encryption and technologies protect the data from the hacker and the full encryption key lifecycle management process establishes a powerful security system which can increase IoT security. In this context blockchaining can be considered one of the best approaches to establishing security.

9.5 SECURITY ANALYTICS AS A DIMENSION OF IOT SECURITY SOLUTIONS

To facilitate the necessary measures in connection with the Internet of Things security, we need to monitor all the smart devices and carry out regular monitoring and check their performance of the security standard. For this purpose various IoT security vendors offer their analytical capabilities and the tools. For an authentic solution, vendors assist in collecting, monitoring, processing and reporting on data that is given to the IoT devices. And to address the issue properly, analytics toolkits are frequently used which can match recent trending technologies, which also include AI, machine learning, deep learning and big data. In general security analytics can be considered an integral part of the software solution for security analytics.

9.6 IOT SECURITY TECHNOLOGIES: CORE PROTECTION METHODS (IOT API)

Application program interface (API) security capability includes the ability to authenticate and authorize the flow of information inside the fully protected IoT network, which may include smart devices, backend systems or any third-party applications. The API must provide end to end service which helps the client to proceed with the required platform [13] and the deserved API control ensuring the security of the information and the data. This should guarantee safe storage of sensitive information, and should have a system of authentication and authorization.

9.7 HOW TO BUILD TRUST IN IOT

To build trust in IoT environment we have to ensure the following:

9.7.1 Enable Device Authentication

To ensure secure participation in the Internet of Things, each and every connected device needs a unique identification in the network. We have a number of methods to prove an identity to the IoT devices such as passwords, biometrics, digital certificates etc. However, when we focus on providing the secure ID the choice of device depends on the capabilities of the device [9].

In environments where our main point of consideration is only security and safety, a hardware-based authentication provides the best means to establish and maintain authentication of the device identity. The digital certificates issued from a trustworthy [5] vendor of key management can be the proven mechanism for security, whereas the storage and processing of the information received demands the traditional RSA keys and management by elliptic curve cryptography (ECC). The combination of both RSA and ECC makes device authentication more reliable.

9.7.2 Encryption to Protect Data

With the use of a proper authentication device a proper encryption mechanism is essential to protect the IoT data which is confidential and has high significance. In IoT the device must protect user data from the device itself and the environment and cloud storage.

This requires process steps to identify the data to be encrypted, and also a key management scheme to distribute and manage the keys that are used to encrypt the data. The secure storage and access control for keys need proper investigation and planning before giving permission or the key to the user. Keys need to be properly managed and integrated in order to ensure security. The IoT and its applications are increasing day by day so for more authentic use the key management must be scalable and dynamic all the time.

9.8 BLOCKCHAIN TECHNOLOGY

A blockchain technology based system is a classical distributed system [11] where all the participating entities are geographically scattered but connected through different types of networks [2]. Decentralization is the fundamental characteristic of blockchain which can be employed to solve the issues of traditional transaction management systems. Blockchain basically provides a platform where multiple entities which do not trust each other can work or share information in a common platform.

This technology is a decentralized [12] architecture having all transactions recorded as a digital ledger. All the nodes are connected in a distributed manner such as mesh topology. Transactions occurring between any nodes are passed through verification by the blockchain network, then after the mining process the completed transaction is recorded in a block. The block consists of numbers of valid transactions between

TABLE 9.1
Comparison of permissionless and permissioned blockchain

	Permissionless Blockchain	Permissioned Blockchain
Environment Type	Open	Closed
Participation in Consensus	All nodes	Selected nodes
Identity	Pseudo-Anonymous	Registered Participants
Consensus Type	Lottery based	Voting based
Transaction Processing Speed	Slow	Fast
Consensus Algorithms	Proof of Work, Proof of Stake, Proof of Burn, Proof of Delegated Stake, etc.	Paxos, Practical Byzantine Fault tolerance, Raft, etc.

different nodes. Once recorded in a block the transaction can never be changed. So blockchain provides an immutable digital ledger among all the nodes present in the network. It builds trust [7] among all users as all the users have the same set of digital record present among themselves and whatever happens in the network one can see it. Blockchain technology has security, and privacy issues of the users are addressed using public and private key concepts also using a digital signature. Blockchain can mainly be used in two different ways; permissionless and permissioned. Table 9.1 provides a brief comparison of both the technologies. Permissionless design which is generally established on an open environment allows anyone to join the system as well as allowing writing to shared blocks. Permissionless design also gives equal privilege to all the nodes in case of consensus process. A permissioned blockchain design is managed by a known set of entities and is established in a closed environment. Though all the entities are allowed to perform transactions, only a fixed set of predetermined nodes can take part in the consensus process in a permissioned blockchain. Consensus algorithms [4] play an important part in managing an efficient and secure blockchain system.

9.9 DISTRIBUTED CONSENSUS ALGORITHMS

The concept of consensus is an engrossing topic in a decentralized or distributed network. Consensus means a procedure to arrive at a common agreement in a decentralized or distributed multi-agent platform. In a conventional distributed system, we apply consensus to ensure reliability which ensures correct execution in the presence of faulty individuals and fault tolerance. In addition to this, a distributed consensus mechanism should satisfy certain properties such as termination, validation, integrity and agreement. An example of consensus is state machine replication, which is a key aspect of any distributed consensus protocol. For example, if we want to run some kind of distributed protocol over a network, every individual entity runs the current protocol and they store the state of the protocol in different state machines. So the entire execution part of the protocol can be represented as a state machine. Now this state machine needs to be replicated to multiple entities so that every individual entity can reach a common output of the protocol. Achieving consensus can be easy and straightforward for certain architectures under certain scenarios. The scenarios could be either that the entire system is faultless or that there is not be any failure in the system, so that every entity can receive the message [19] correctly or the system behaves in a synchronous way, i.e., it is expected that you will receive all messages within some predefined time interval.

However, achieving consensus can be non-trivial in the case of a distributed environment [13] due to the presence of multiple types of failure. Typically in a distributed system, we consider three different types of failure:

1) Node Failure: A node suddenly crashes or becomes unavailable in the middle of communication. Therefore we are not expected to receive any message from that particular node. This can be a hardware or software fault.
2) Partitioned Faults: This type of fault occurs whenever links fail, which results in partition in the network. This can hamper reaching of consensus.
3) Byzantine Faults: This kind of fault is more difficult to handle in a distributed environment. Here the entity starts behaving maliciously. In both the above faults, we can expect an effect on the network, but in this case, it is difficult to guess because it completely depends on how maliciously the entity is behaving and what the entity is doing.

The correctness of a distributed consensus protocol can be characterized by the following two properties; safety and liveliness. Safety ensures that one will never converge to an incorrect state. And liveliness ensures every correct value must be accepted eventually.

9.10 CONSENSUS IN PERMISSIONLESS BLOCKCHAIN SYSTEM

Conventional consensus mechanisms will not be applicable in an open or permissionless blockchain system. Therefore the following section shows how consensus has been achieved in bitcoin-like open environments along with their shortcomings.

9.10.1 Bitcoin Consensus

The main purpose of consensus in bitcoin is to add a new block to the existing blockchain. There can be multiple miners in the bitcoin network and these miners can propose their new blocks based on the transactions they have heard of. It is not necessary that every miner propose the same block. Generally the miners include those transactions in the new block that they have heard of since the last time a block has been added. The miners also need to ensure that size of the newly proposed block does not exceed a certain threshold. We should focus on the following two observations before designing the algorithm:

Any valid block (a block with all valid transactions) can be accepted even if it is proposed by only one miner.

The entire protocol can work in rounds. Broadcast the accepted block to the peers and collect the next set of transactions.

Based on the above two observations, we can design solutions so that every miner independently tries to solve a problem and the block is accepted for the miner who can prove first that the challenge has been solved.

9.10.2 Proof of Work (PoW)

The idea of PoW came in the year 1992 from Dwork and Naor to combat junk emails where we have to do some work to send a valid email. The attacker this way will be discouraged from sending junk emails because in that case they have to do some work of some complexity in order to forward the junk emails that does not prove beneficial for them. A blockchain-based PoW system must have certain features such as asymmetry: The task must be relatively hard but feasible for the service requester; the task must be easy to verify for the service provider. In this way the service requester will get discouraged from carrying out the work but the service provider can easily check the validity of the work because of the asymmetric nature of the work. Bitcoin-based PoW systems extend the hashcash-based PoW system and develop a methodology to protect the blockchain by applying the distributed consensus mechanism. The hashcash system was proposed by Adam Back and uses the puzzle-friendliness property of the cryptographic hash function. Now coming to bitcoin-based PoW, miners who take part in the consensus process need to give a proof that they have done some work before proposing a new block. The attacker will be discouraged from proposing a new block or making change in existing blocks. Because in that case they have to do the entire work of the blockchain which is computationally difficult in a generic environment. Every miner will try to find out a nonce value which will satisfy a certain hash equation. Most implementations of bitcoin PoW use a double SHA256 hash function. In the default set up, all miners wait for around ten minutes and look for all transactions which have taken place within that duration. Then they start the mining process. As we have seen, the probability of getting a PoW is very low; hence it will never happen that any miner will be able to control the bitcoin network exclusively. In case any attacker wants to make some changes in one block, they have to do more work compared to the collective work of all the blocks in the longest chain, which is computationally difficult, though not impossible, thus making an attack difficult with current hardware and hence the bitcoin system is tamper-proof. These tamper-proof characteristics of PoW also ensure that double spending does not happen in the case of a blockchain network. Apart from this, PoW-based systems suffer from a number of security attacks such as sybil attack, DOS attack, etc. In the case of a sybil attack, the attacker tries to fill the network with clients under its control. This helps the attacker to control the entire network. These clients work according to the instructions of the attacker which compromises the PoW mechanism. In DOS, the malicious node will send a lot of data to a particular node or nodes which will hamper the normal functioning of the bitcoin transactions. Another major flaw in a bitcoin system is the monopoly problem. This happens when a particular miner gains control of the network by deploying huge servers for the mining process. So it may happen that this particular miner will gradually generate a huge number of blocks and hence the miner can control the entire flow of transactions. As a result, other nodes will get discouraged from joining as miners and only a few miners with large computing resources [10] will control the network.

9.10.3 Proof of Stake (PoS)

The PoS mechanism is an improved version of PoW to reduce the excessive electricity consumption of PoW-based systems. As a substitute to the computationally expensive PoW mechanism, PoS aims to stake nodes' economic share in the network. Like PoW, a node (miner) is selected to add a block to the blockchain. But here the miner selection procedure is different from PoW. In PoS, the selection of a miner is proportional to the amount of bitcoin it holds. Block finality in the PoS-based system is faster compared to PoW blockchain, as computationally difficult puzzle solving is not within PoS. The PoS-based algorithm developed by Ethereum is known as Casper. The PoS algorithm pseudo-randomly chooses miners to create blocks of the blockchain, so that no miner can predict its turn in advance. It also solves the monopoly problem of PoW-based consensus. Naive PoS algorithms are prone to an attack known as Nothing-at-Stake, and thus require some improvements in order to provide safety.

9.11 CONSENSUS IN PERMISSIONED BLOCKCHAIN SYSTEM

Permissioned blockchain consensus is achieved through the help of a smart contract, which is basically an extension of bitcoin scripts. Smart contracts are self-executing software programs in which the terms and conditions among the negotiating parties are written down. Generally, the concept of state machine replication mechanism is used to ensure consensus in the permissioned blockchain environment because of the following reasons such as the network being closed and the nodes knowing each other, so state replication is possible among the known nodes. And it avoids the overhead of mining.

9.11.1 Paxos

The first ever consensus algorithm, Paxos, was proposed by Lamport with the objective of choosing a single value under crash or network fault. The idea behind Paxos is very simple: All the nodes in the network have been categorized into three types; namely the proposers, the acceptors and the learners. Everyone is a learner in the network that learns the consensus value. The proposer initially prepares a proposal with a proposal number known as a prepare message and sends it to the acceptors. This proposal number forms a timeline and the biggest number is considered up to date. Each acceptor compares received proposal number with current known values for each proposer's proposal message. If it gets a higher number, then it accepts the proposal or otherwise declines it. Then the acceptor prepares a response message of the following form: Prepare response where the proposal number is the biggest number the acceptor has seen and accepted values are the already accepted values from another proposer. Next a vote is taken based on the majority decision. The proposer checks whether the majority of the acceptors have rejected the proposal. If yes, then the proposer updates it with the latest proposal number. If no, then the proposer further checks whether the majority of the acceptors have already accepted values. If yes then the proposer's value cannot be selected or otherwise it sends an accept message. Finally, the proposer sends an accept message with the following format to all the acceptors. Whenever the acceptor accepts a value, it informs the learner nodes about it so that everyone will learn about the accepted value. If more than (Nonce/2 – 1) acceptors fail, then no proposer will get enough reply messages to form a conclusion and we cannot reach consensus. Even though the concept behind Paxos is very simple to understand, the theoretical proof is very complicated. Real-life application of this may need to go for a sequence of selections which is known as a multi-Paxos system. Multi-Paxos systems need a repeated application of Paxos which increases the system complexity. This is because the number of messages exchanged between the nodes also increases. There is also a chance of livelock or starvation in Paxos-based systems.

9.11.2 Raft

The Raft algorithm is designed as an easy alternative to Paxos. The basic idea behind Raft is that nodes collectively select a leader and rest of the nodes become followers. The leader is responsible for state transition log replication across the followers. Record entries flow in only one direction from the leader to the followers. Unlike Paxos, here each node can be at any of the three states namely leader, candidate and follower at any particular time. The Raft algorithm runs in rounds which are known as term. Each term starts with an election where one or more candidates strive to become leader. Initially, we have a set of follower nodes, who look out for a leader. If within a certain time interval they do not find one, then the leader election process starts. In this election phase, some of the followers volunteer to become a leader and request votes from all other nodes. Then these candidate nodes send request messages to other followers of the system for a vote. When a node receives the request, it compares the term and index in the received message with corresponding current known values. Like Paxos, the followers vote for one of the candidates and based on the majority of votes, a leader is selected. Then in the next part, the elected leader will propose values which the followers will choose so that the system can reach consensus. The leader sends out heartbeats (signals) to all followers at regular intervals in order to maintain its authority. Compared with Paxos and PBFT, this algorithm has high efficiency and clarity. Hence it has been extensively employed in distributed systems. The Raft algorithm achieves the same safety performance as Paxos and is better suited in real-life implementation and comprehension. As mentioned, the Raft algorithm cannot support byzantine nodes and can stand up to failure of 50% of nodes. As in the case of permissioned blockchain, nodes are verified members. Hence, it is more essential to resolve crash faults rather than Byzantine faults for private blockchain.

9.11.3 Byzantine Fault Tolerance and Its Variant

In relation to distributed systems, Byzantine Fault Tolerance is the capability of a distributed network to execute as required and correctly reach a consensus despite malicious nodes. It is derived from the Byzantine General Problem, where the general sends an attack message to one group of lieutenants whereas he sends a retreat message to another group of lieutenants. Hence it becomes difficult for the system to find out what action to take. Byzantine fault-tolerant systems are typically built using replication. For this, the state machine approach is used which helps to implement fault-tolerant services. The variant of BFT that has been designed for synchronous distributed systems is called the "Lamport–Shostak–Pease" algorithm. This ensures consensus in presence of a number of faulty nodes, provided we have (2f + 1) number of lieutenants apart from the commander. But our real systems behave in an asynchronous way as there is no guarantee that a message will be received within a certain time interval. For this reason, a variant of BFT known as Practical Byzantine Fault Tolerance (PBFT), has been developed for real-life asynchronous systems. As in the case of a pure asynchronous system, achieving consensus is impossible even in the presence of a single faulty node. So to ensure liveness property, instead of pure asynchronous system, a weak asynchronous system has been considered. Coming to the algorithm, the byzantine model consists of three types of nodes: the clients, a commander and the lieutenants. The entire algorithm runs in three phases: pre-prepare, prepare and commit phase. In the pre-prepare phase the commander assigns a sequence number to the request submitted by a client and multicasts it to the network. Among other data, the request message also contains the digital signature and message digest for verification. The lieutenants of the network confirm the block by verifying the digital signature and message digest. Once the validating lieutenants accept the pre-prepare message, they enter the prepare phase by multicasting the message to the rest of the network. Once again both the commander and lieutenants verify the prepare messages before accepting them. The messages commit, when (2f) prepare message from different backups match with the corresponding pre-prepare messages. Hence, the total (2f + 1) votes one from primary from the non-faulty replica help the system to reach to a consensus. With the evolution of ICT, the blockchain technology has attracted interest from various directions. The consensus algorithm is the main technology of blockchain. In the case of permissionless systems, it is easy to achieve robust consensus among large number of untrusted nodes using complex computations though transaction, finally remains non-deterministic. On the contrary, permissioned blockchain provides high throughput in less time while sacrificing a degree of decentralization.

Since the concept of IoT arose, much development has taken place in a different application. The architecture of IoT is basically three layer perception, network, and application layer to make real use of the IoT technology for efficiency and reliability of the system. Security and privacy issues are the challenges [20] to implement a different application using the IoT concept. Some of the challenges are access control, authentication, centralized or distributed network, and the identity of the things. Visualization is one the important aspects [21] of the IoT application. The end-user monitors the environment using either mobile phone or smart tablets, although IoT applications use both centralized and decentralized architecture. The decentralized architecture has an advantage over centralized architecture. The IoT application system has different stages. Processing and doing computation may depend on cloud computing or the recent development of fog computing. The storage purpose IoT application can store at the edge device or at the cloud server depending upon the storage space availability. In IoT / physical layer all the required sensors are deployed in the application area. The sensors are connected to the next layer using WiFi or through a wire. IoT application demands faster processing and quick response for better utilization of the system. Most of the application collecting data from sensors need to process in the edge of the network for faster processing and temporary storage for quick analysis of the data. The decision can be taken by analyzing the data either in a collaborative way or a distributed way by mutual agreement. All these activities can be done in the Fog Layer / Processing layer. The Cloud Layer provides different types of services such as infrastructure, platform, and software as a service. In an IoT application as different sensors collect information from environment continuous way, it is not possible to provide large memory space. Also, IoT devices are low in memory capacity. Similarly, the storage capacity of a fog device is also limited, so the final storage must be provided by cloud computing. For establishing security and privacy in communication and user data, authentication of IoT devices is highly essential. When devices perform any operation in an application, the devices need to be identified using their unique ID. Using the unique ID, devices can connect to the next layer also performing different computation in a collective way. The authentication of the devices means the communication can be done securely; all the nodes are identified using their ID. The decentralized scheme for IoT device authentication can be considered one of the best approaches for secure communication. The process of device authentication is done in blockchain-based technology. Implementation takes place using the Ethereum platform and Web3 client is used to integrate block chain smart contract with Frontend.

9.11.4 Security Analysis

IoT implementation requires many security and privacy issues to be addressed. Here a decentralized approach is proposed for authentication to increase IoT network connectivity and build trust among all the devices. In an IoT system not only information needs to communicate securely it also needs to identify which devices are authenticated devices. Each of these devices identify using their public key which is a unique key generated from the system. Security and integrity of the IoT system are ensured without using a centralized system. The proposed authentication protocol [17] is secured against some of the attacks described below:

1) **MITM (Man-In-The-Middle)**: In the proposed scheme, MITM attack is not possible as whatever message is sent from one node to another node, it is sent using hash techniques and public-private key concept so that only the authorized user will get the message. Any modification done on the message will be ignored by the receiver side. The message communication also uses a digital signature as one of the concepts of blockchain technology.
2) **Impersonation Attack:** Whatever transaction is done in the blockchain network, all are verified and mined by the network using mining process and digital signature concept. Each user has their own identity added to the system using the authentication process.

3) **Replay Attack:** In a blockchain environment, no node can capture more than half of the network power. Each transaction sent or received is recorded in a digital ledger after verification and mining. So no transaction can transmit multiple times in the network.
4) **Denial-of-Service (DoS) Attack:** As each message is broadcast in the network and it goes through a verification process to check valid or invalid transaction, there is no chance that a DoS attack is possible.

Due to all the above security issues, a decentralized web authentication system using blockchain is proposed which is not a password-based authentication and authentication is done using AuthKey which is a 160-bit hash and is secured enough to prevent all the above attacks.

9.11.5 Decentralized Authentication

Decentralized authentication is an attempt to make a decentralized site login and authentication protocol. It is analogous to the "Log In with Facebook" button that we have probably become accustomed to. It is a smart contract that will store user IDs and their associated wallet addresses. The user ID is simply a UTF-8 string with size ranging between 2 and 32 bytes. The user himself creates it on inception of the wallet and will later use it to enter any site that supports decentralized authentication. It would also be possible to add a restriction on the possible characters included in the string. One could restrict it to Latin characters and Arabic numerals in order to limit the possibility of creating visually similar IDs. When creating an account with decentralized authentication, a pair of keys is created. We will create an authorization key and a key to restore access. When created, both addresses are the same as the address of the wallet which first made the transaction. Users who care about their security should create a separate Master Key and store it in a place that is inaccessible from online. Recovery seeds are a set of 12 mnemonic words that when used can recreate the key pairs for our wallet. If we are going to be using a wallet for authentication then it is also recommended we use a separate address from the one that keeps all of our ether. Doing this allows us to avoid any hackers from tracing our Authkey to the wallet with our assets. Decentralized Web Authentication System using Ethereum-based blockchain protects our assets.This is something that could be updated in later iterations of the smart contract. If we want to further protect ourselves, then we consider VPN services.

9.11.6 Using Decentralized Authentication

There is a dedicated web page intended for user interaction with the smart contract. We can create an account there, change the keys or delete it. To work with it, the user will need to install the well-known browser plugin called MetaMask. Of course, if we are already an experienced user of the Ethereum network then we will already have used MetaMask and will probably have an idea of how it interacts with the network. The overall user authentication process using DecAuth looks as follows: The site (backend) contacts the smart contract and receives the user's Ethereum address.

Then the site (backend) generates and records a message, and asks the user to sign this message with the help of the authKey address. And the user being on the site (frontend) signs the message using the MetaMask plugin and sends it to the backend. Finally the site (backend) verifies the signature, and if everything is correct, it activates the user's session. It is important that authentication checks should take place in a user-uncontrolled environment. So, in other words, all of the checks should be completed on a server instead of on a user's browser.

The decentralized system is not subject to censorship by the large entities, such as Google or Facebook. If it is necessary to censor something, each website should implement it independently. Yet this would only affect the user's interaction with that site and not any others. The Ethereum network currently has quite slow transaction speeds when creating an account; the user may have to wait a few minutes, but sites can get the data and verify users quite quickly. This solution scales well, because there are a lot of data nodes, and anyone can add another one at any time. The complexity of implementing such a solution for site owners is no higher than the complexity of implementing OAuth 2.0. Using blockchain-based authentication, all devices are connected in a peer-to-peer way. Each of these devices identifies using their public key which is a unique key generated from the system. Security and integrity of the IoT system is ensured without using a centralized system. Due to different kinds of security issues, a decentralized web authentication system using blockchain is proposed which is not a password-based authentication and authentication is done using AuthKey which is a 160-bit hash and is secure enough to prevent all the above attacks and the blockchaining method makes it secure against all kinds of attack.

REFERENCES

[1] Mahmoud Ammar, Giovanni Russello, Bruno Crispo (2018) Internet of Things: A Survey on the Security of IoT Frameworks, Journal of Information Security and Applications (JISA), vol. 38, pp. 8–27.

[2] Sachi Nandan Mohanty, K.C. Ramya, S. Sheeba Rani, Deepak Gupta, K. Shankar, S.K. Lakshmanaprabu, Ashish Khanna (2020) An Efficient Lightweight Integrated Blockchain (ELIB) Model for IoT Security and Privacy, Future Generation Computer Systems, vol. 102, pp. 1027–1037.

[3] Bhabendu K. Mohanta, Anisha Sahoo, Shibasis Patel, Soumyashree S. Panda, Debasish Jena, Debasis Gountia (2019) "DecAuth: Decentralized Authentication Scheme for IoT Device Using Ethereum Blockchain", in Proc. of the 2019 IEEE Region 10 Conference (TENCON), pp. 1–6

[4] Soumyashree S Panda, Bhabendu Ku. Mohanta, Utkalika Satapathy, Debashis Jena, Debasis Gountia, Tapas Kumar Patra (2019) "Security (In-Depth) Analysis of Blockchain Based Decentralized Consensus Algorithms", in Proc. of the 2019 IEEE Region 10 Conference (TENCON), pp. 1–6

[5] Bhabendu K. Mohanta, Utkalika Satapathy, Soumyashree S. Panda, Debasish Jena, Debasis Gountia (2019) "Trustworthy Management in Decentralized IoT Application using Blockchain", in Proc. of the IEEE International Conference on Computing, Communication and Networking Technologies (ICCCNT), pp. 1–5

[6] Soumyashree S. Panda, Utkalika Satapathy, Bhabendu K. Mohanta, Debasish Jena, Debasis Gountia (2019) "A Blockchain Based Decentralized Authentication Framework for Resource Constrained IOT Devices", in Proc. of the IEEE International

Conference on Computing, Communication and Networking Technologies (ICCCNT), pp. 1–7.

[7] Sicari, Sabrina, Alessandra Rizzardi, Luigi Alfredo Grieco, and Alberto Coen-Porisini (2015) "Security, Privacy and Trust in Internet of Things: The Road Ahead." Computer Networks, vol. 76, pp. 146–164.

[8] Zhang, PeiYun, MengChu Zhou, and Giancarlo Fortino (2018) "Security and Trust issues in Fog Computing: A survey." Future Generation Computer Systems, vol. 88, pp. 16–27.

[9] Kang, Kai, Zhibo Pang, Li Da Xu, Liya Ma, and Cong Wang. "An interactive trust model for application market of the internet of things." IEEE Transactions on Industrial Informatics 10, no. 2 (2014): 1516–1526.

[10] Jeong, Seohyeon, Woongsoo Na, Joongheon Kim, and Sungrae Cho. "Internet of things for smart manufacturing system: Trust issues in resource allocation." IEEE Internet of Things Journal 5, no. 6 (2018): 4418–4427.

[11] Roman, Rodrigo, Jianying Zhou, and Javier Lopez. "On the features and challenges of security and privacy in distributed internet of things." Computer Networks 57, no. 10 (2013): 2266–2279.

[12] T. Salman, M. Zolanvari, A. Erbad, R. Jain, and M. Samaka, "Security services using blockchains: A state of the art survey," IEEE Communications Surveys & Tutorials, vol. 21, no. 1, pp. 858–880, 2018.

[13] T. Neudecker and H. Hartenstein, "Network layer aspects of permissionless blockchains," IEEE Communications Surveys & Tutorials, vol. 21, no. 1, pp. 838–857, 2018.

[14] J. Chauhan, S. Seneviratne, Y. Hu, A. Misra, A. Seneviratne, and Y. Lee, "Breathing-based authentication on resource-constrained iot devices using recurrent neural networks," Computer, vol. 51, no. 5, pp. 60–67, 2018.

[15] P. Gope and B. Sikdar, "Lightweight and privacy-preserving two-factor authentication scheme for iot devices," IEEE Internet of Things Journal, vol. 6, no. 1, pp. 580–589, 2019.

[16] F. Sun, C. Mao, X. Fan, and Y. Li, "Accelerometer-based speed-adaptive gait authentication method for wearable iot devices," IEEE Internet of Things Journal, vol. 6, no. 1, pp. 820–830, 2019.

[17] Z. Wang, "A privacy-preserving and accountable authentication protocol for IoT end-devices with weaker identity," Future Generation Computer Systems, vol. 82, pp. 342–348, 2018.

[18] J. Gubbi, R. Buyya, S. Marusic, and M. Palaniswami, "Internet of Things (IoT): A vision, architectural elements, and future directions," Future generation computer systems, vol. 29, no. 7, pp. 1645–1660, 2013.

[19] R. Roman, J. Zhou, and J. Lopez, "On the features and challenges of security and privacy in distributed Internet of Things," Computer Networks, vol. 57, no. 10, pp. 2266–2279, 2013.

[20] M. A. Khan and K. Salah, "Iot security: Review, blockchain solutions, and open challenges," Future Generation Computer Systems, vol. 82, pp. 395–411, 2018.

[21] I. Makhdoom, M. Abolhasan, H. Abbas, and W. Ni, "Blockchain's adoption in iot: The challenges, and a way forward," Journal of Network and Computer Applications, 2018.

10 Geospatial Data Classification using Sequential Pattern Mining with Modified Deep Learning Architecture

Sunil Kumar Sahoo and Brojo Kishore Mishra
GIET University, Gunupur, India

CONTENTS

10.1 INTRODUCTION

Advances in geographic information systems (GIS) provide an opportunity to gather, store, edit, query, verify, share and manifest geographically referenced information. In the case of earth science, a huge amount of data has been collected at various levels of granularity and this geospatial data is always big data [9][10]. At present, analysis of geospatial big data allows users to scrutinize huge amounts of geospatial data as this geospatial big data is spatial data sets that are beyond the capacity of current computing systems. For instance, the National Oceanic and Atmospheric Administration's (NOAA) satellites gather ocean, coast and atmospheric data of the global ecosystem for understanding and predicting changes in the Earth's environment [11]. The growth of this spatial data leads researchers to conduct further researches on spatial data mining techniques in a highly automated fashion.

Spatial data mining (SDM) is utilized for the extracting and mining of hidden, implicit, valid, novel and interesting spatial or non-spatial patterns or rules from large-amount, incomplete, noisy, fuzzy, random, and practical spatial databases [12] [13]. The efficiency of the SDM is based on the mining algorithm only. Moreover, one of the issues related to the SDM is that the geological data is typical spatial data, which includes geological, geophysical, geochemical, and remote sensing [14]. Big data is still not a clearly defined term and it has been defined differently from technological, industrial, research or academic perspectives. So, for solving these issues, the big data is classified as structured and unstructured datasets with massive data volumes that cannot be easily captured, stored, manipulated, analyzed, managed or presented by traditional hardware, software or database technologies since big data is often described by its unique characteristics [15].

Geospatial big data can be characterized by the following [22]: (a) Volume: Records sensed imagery data in Petabytes. Increase in data sets produces massive issues in storage (b) Variety: Includes the map data, imagery data, geo-tagged text data, structured and unstructured data, raster and vector data, (c) Velocity: imagery data with frequent revisits at high resolution, continuous streaming of sensor observations, Internet of Things (IoT), real-time GNSS trajectory and social media data all require matching the speed of data generation and the speed of data processing to meet demand (d) Veracity: much of geospatial big data is from unverified sources with low or unknown accuracy; the level of accuracy varies depending on data sources, raising issues about quality assessment of source data and how to "statistically" improve the quality of analysis results. (e) Visualization: This helps analysts identifying patterns (such as outliers and clusters), leading to new hypotheses as well as efficient ways to partition the data for further computational analysis. (f) Visibility: the emergence of cloud computing and cloud storage has now made it possible to efficiently access and process geospatial big data in ways that were not previously possible [16][17]. The variation in the format of data collection and increase in the volume of data are a challenging issue to geospatial data processing as it creates issues in storing, managing, processing, analyzing, visualizing and verifying the quality of data [18][21]. Moreover, the verification in the quality of geospatial big data and data products delivered to end-users is another challenging issue in quality control. On the other hand, fitness of uses or purposes appears more valid or should be advocated in the context of big data [19][20].

10.2 RELATED WORKS

In 2015, Hillen *et al.* [1] proposed the Geo-reCAPTCHA (Completely Automated Public Turing test to tell Computers and Humans Apart) in order to gather user-generated geographic information (UGCI) from earth observation data, as the UGCI was utilized for geography and geographic information science (GIScience). The Geo-reCAPTCHA approach was mainly based on assessing the time and quality of the resulting geographic information. Most of the problems related to the building digitization were solved within a short amount of time (19.2 s on average) and the accuracy of digitization in geospatial data was improved to 82.2%. Moreover, Geo-reCAPTCHA had high performance when compared to the reCAPTCHA. Thus, Geo-reCAPTCHA was a data-rich channel belonging to crowd-sourced geographic information.

In 2015, Kebler *et al.* [2] proffered the time geography for the purpose of querying and integrating multiple spatial–temporal data sources and this framework was designed under the perception of the space–time prism and this was based on the constraints and interactions between entities in space and time. This research primarily developed the space–time prisms and then applied it to the Web of Data to get efficient spatial–temporal aspects. This research mainly focused on monitoring the environment with the space–time prisms and thus the spatial–temporal and semantic reasoning was obtained with the geospatial information and this information was gathered directly from distributed data sources.

In 2018, Xia *et al.* [3] accessed three earth observations (EO), namely fast access, accurate service estimation and global access with the help of accessing big data. Moreover, the research explored the spatial pattern, temporal pattern and spatio-temporal pattern of user-data interactions. With the help of these patterns, geospatial information was gathered by overcoming the drawbacks associated with the spatiotemporal patterns when end-users access EO data. This research also solved the problems associated with the utilization of spatiotemporal patterns for facilitating better EO big data access with the help of three spatiotemporal optimization strategies; they are (i) spatiotemporal indexing to accelerate data access, (ii) spatio-temporal service modeling to improve data access accuracy and (iii) spatiotemporal cloud computing to enhance global access. This research had a better framework in the optimization of EO big data access and was also vulnerable to other multidisciplinary geographic data and information research.

In 2018, Lu *et al.* [4] projected the GreenBDT: Renewable-aware scheduling of bulk data transfers for geo-distributed sustainable datacenters for the purpose of minimization of grid energy cost for bulk data transfers between sustainable and green datacenters and for maximizing the usage of renewable energy. GreenBDT was made possible with the heuristic algorithm in order to solve the MaxGreen-Min-Cost problem. This algorithm is employed with the following constraints; in case available wind power was not enough for the transfer all the bulk data, then the wind energy was utilized for the interGD BDTs or in case of insufficient wind power, then the inter-GD BDTs used optimal demand division and routing selection to minimize the energy cost caused by grid power. This research further revealed that validation with the real-life network topology optimized the scheduling for inter-datacenter bulk data transfers individually. This further suggested that the renewable energy was maximized. More securitized results proved that with real-life network topology, existing wind power and electricity prices, with maximum renewable energy, reduced the cost of energy savings when compared to the existing bulk data transfer strategies

In 2011, Jiamthapthaksin *et al.* [5] formulated the generic agglomerative clustering framework for the purpose of generalizing agglomerative clustering in the geo-referenced datasets (GAC-GEO) by using three plug-in components. The fitness function of the plug-in is maximized with the GAC-GEO agglomerates as it had the capacity to capture the notion of interestingness of clusters. This further enhanced the typical agglomerative clustering algorithms by using fitness functions support task-specific clustering, whereas generic neighboring relationships increase the number of merging candidates. This shows that the existing agglomerative clustering algorithms were considered as the specific cases of GAC-GEO. The proposed framework was evaluated on an artificial dataset and two real-world applications involving region discovery.

In 2017, Goel *et al.* [6] presented a dual algorithm; one was a hybridized version of two nature-inspired algorithms such as the Bat algorithm and Charged System for the purpose of better classification of the images and the second algorithm was the Clonal Selection Algorithm for Search for the purpose of feature extraction of geospatial big data as the geospatial feature extraction was crucial for remote sensing. Furthermore, the data set became more complex because of the usage of multispectral satellite images; this complexity further increased the number of dimensions for the classification problem. Moreover, this suggested that the proposed nature-inspired algorithm was good at classification of the land covers of satellite images. The evaluation of the efficiency of the proposed algorithms was made with the dataset of multispectral satellite images in order to prove that the proposed algorithm was better than the existing one.

In 2017, Shi *et al.* [7] suggested the affinity propagation (AP) algorithm as a novel machine learning algorithm for the purpose of classification of geospatial images related to the earth observations. The AP had a restriction access in handling large data. Moreover, the serial computer program took a long time to complete the AP calculation. Therefore multi-core and many-core computer architectures were accompanied by application accelerators by parallel combination of developing tasks and data levels; this was done for the purpose of guaranteeing and achieving scalable geocomputation. Moreover, for spatial cluster analysis, this AP algorithm was parallelized with the processing unit (GPU). This parallelization produced an optimal solution for the issues of big geospatial data processing and its broader impact for the GI Science community.

In 2018, Barik *et al.* [8] proposed a mist computing-based framework for the purpose of data mining and analyzing of the geospatial big data. The prototype here was built by utilizing the Raspberry Pi, an embedded microprocessor. The testimony of the developed *MistGIS* framework was carried by the preliminary analysis including K-means clustering and overlay analysis. The outcomes proved that mist computing had the capacity to assist fog and cloud computing and it too guaranteed the analysis of big data in geospatial applications.

10.3 PROBLEM DEFINITION

The literature has come out with several techniques for the EO based on geospatial data processing as shown in Table 10.1. However, they require more improvements because of lack of several features in data processing. Geo-reCAPTCHA [1] had generated geographic information from earth observation data and had the eligibility to distinguish between a human and a machine in case of data processing. It had reduced spam and viruses. The major drawbacks here were high data errors, low reliability and low quality. Moreover in a few cases it is very difficult to read data from clusters and requires more time to decipher. In case of the space–time prism model [2], common ontological problems were solved by template solutions. It had less spatial autocorrelation that essentially quantified the correlation parameters within itself through space and the spatial autocorrelation was positive, negative or zero to improve the quality of processing the data. The major challenges that need to be overcome here are less accessibility, less convergence and limitations in speed. With keyword-based matching algorithms [3] fast access and accurate service estimation was available; it also had pros as utilization of data demand patterns in different regions and time windows, and had the capacity to monitor EO data services from different geolocations. It had introduced the cloud-computing capabilities to facilitate EO data access and also optimized a cloud-computing data framework with spatiotemporal data access patterns. The cons of this method are that it needs deep learning technologies to improve access pattern prediction capabilities, had limitations in memory and resource allocation, more complex to cloud-computing infrastructures and the selection of appropriate cloud-computing services has become a barrier to cloud adoption in EO big data science. Moreover, in the case of the time-aware task scheduling algorithm [4] there was maximization in green energy usage and minimization in grid energy cost; it used all the wind power to transfer bulk data at each time slot and had reduction in energy cost. The major drawbacks were that it only focused on work load scheduling in a single green datacenter, and it had a more complicated scheduling and needs an optimal routing and bandwidth allocation for each task. Then, with the agglomerative clustering algorithms and density-based clustering algorithm [5] there was maximization in plug-in fitness function that captured the notion of interestingness of clusters, generic neighboring relationships increased the number of merging candidates; it had provided powerful region discovery capabilities and had demonstrated the capabilities of GAC-GEO in real-world case studies involving an earthquake dataset. It had drawbacks such as; with the round-robin approaches there was no improvement in fitness function, the selection of parameters were made without backtracking, it had no capability to identify arbitrarily shaped clusters, it had had high variance and high correlation with respect to non-spatial attribute(s). Furthermore, in the case of Swarm and artificial immune system-based intelligence techniques [6] it had advantages such as better classification of images, good quality of geospatial feature extraction and the spectral signature of each category is calculated using the training data. It had drawbacks such as high computational complexity and had no capacity to cover the land cover feature. Moreover with the Affinity propagation (AP) algorithm [7] there were advantages with GPU for spatial cluster analysis; it had the capacity to track the data that does not lie in a continuous space and had handled big data in an accurate way. It had challenges as low scalability, memory constraints, AP does not specify a predefined arbitrary number of clusters in advance so that dynamic scheduling was made and it also does not satisfy the triangle inequality. Moreover, with the K-means clustering algorithm [8] there was reduction in the latency period and it had increased throughput, it had adequate storage for data visualization and analysis, it had the capacity of local processing that had led to the reduction in data size, lower latency, high throughput, and power-efficient systems. All these were features of this algorithm and it had challenges in the case of cloud computing as it was not reserved for long-term analysis and it required efficient big data analysis and processing. In order to overcome all these challenges, there is a necessity to propose an optimal solution for geospatial big data processing.

TABLE 10.1

Features and challenges of geospatial big data processing

Author [citation]	Adopted methodology	Features	Challenges
Hillen *et al.* [1]	Geo-reCAPTCHA model	• Distinguished between a human and machine during the problem-solving approach. • Made the online polls more legitimate. • Reduced spam and viruses.	• High data errors • Less reliability and low quality in data processing • Sometimes very difficult to read data from clusters • More time consuming to decipher.
Kebler *et al.* [2]	Space–time prism model	• Common ontology modeling problems were solved • Less spatial autocorrelation • Spatial autocorrelation, was in positive, negative or zero to improve the quality	• Has constants with speed • Less accessibility • Less convergence
Xia *et al.* [3]	Keyword-based matching algorithm	• Fast access • Accurate service estimation • Monitor EO data services from • Different geolocations.	• Had limitations in memory and resource allocation • More complexity in cloud-computing infrastructures
Lu *et al.* [4]	Time-aware task scheduling algorithm	• Maximization of green energy usage and minimization of grid energy cost • Uses all the wind power to transfer the bulk data at each time slot. • Energy saving was high.	• Only focused on work load scheduling in a single green datacenter. • More complicated scheduling • Need optimal routing and bandwidth allocation for each task.
Jiamthapthaksin *et al.* [5]	Agglomerative clustering algorithms and density-based clustering	• Generic neighboring relationships increased the number of merging candidates • Provided powerful region discovery capabilities • High Accuracy	• No improvement in the fitness function • Had no capability to identify arbitrarily shaped clusters • Had high variance and high correlation

TABLE 10.1 CONT.

Author [citation]	Adopted methodology	Features	Challenges
Goel *et al.* [6]	Swarm and artificial immune System-based intelligence techniques	• Better classification of images • Geospatial feature extraction was clear and accurate • The spectral signature of each category was calculated using the training data.	• High computational complexity • Did not cover the land cover feature
Shi *et al.* [7]	Affinity propagation (ap) algorithm	• Graphics processing unit (GPU) was used for spatial cluster analysis. • Had the capacity to track the data that do not lie in a continuous space. • Handled big data in an accurate way.	• Low scalability • Had limited memory space • Did not satisfy the triangle Inequality.
Barik *et al.* [8]	K-means clustering algorithm	• Low latency • High throughput • Had adequate storage for data visualization and analysis • Overhead on cloud server had reduced	• The cloud was not reserved for long-term analysis. • Requires efficient big data Handling and processing

10.4 METHODOLOGY

Figure 10.1 shows the proposed geospatial data classification model. Even though diverse application-related classification models can be developed using geospatial data, the main intention of this research proposal is to know about the kind of industry such as micro industries, macro industries, mid-scale industries, large-scale industries, small-scale and very large-scale industries that suit a specific location. This is because a multitude of factors influence the location decisions of firms and industries, including proximity to raw material supplies, availability of labor, good communications and nearness to markets. Here, geospatial data will be utilized for accomplishing the experiment, from which the X and Y co-ordinate are traced that is related to the latitude and longitude information, respectively. As there is a huge quantity of data it is given to the Hadoop framework for the purpose of handling that big data. Normally, the Hadoop framework is an Apache open-source framework written in java that allows distributed processing of large datasets across clusters of computers using simple programming models. Hadoop framework allows the user to

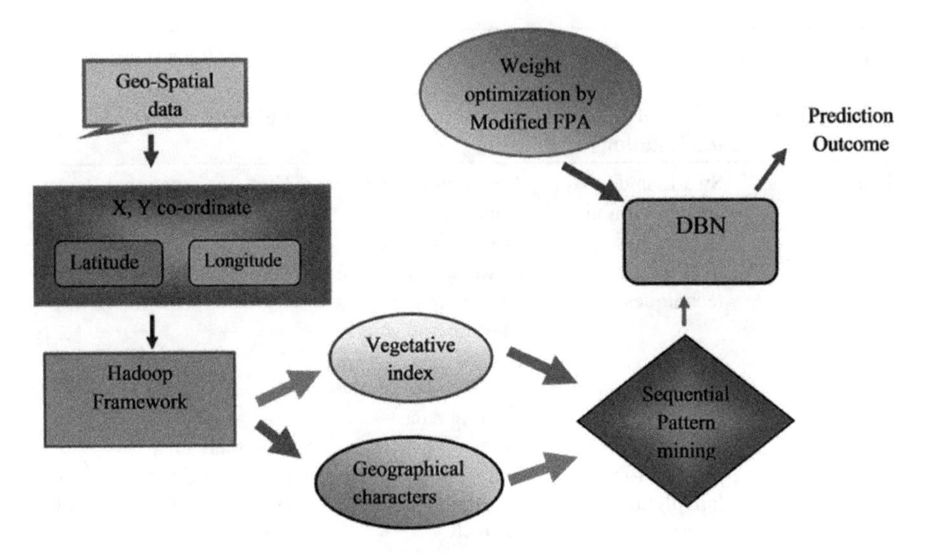

FIGURE 10.1 Block diagram of Geospatial industrial data classification.

quickly write and test distributed systems with big data. Then, from the map reduced data, vegetation indices like simple ratio vegetative index, normalized difference vegetative index, Kauth–Thomas Tasseled cap and Infrared Index Transformation, as well as geographic location will be generated. Next to this data formation, sequential pattern mining will be performed, and those patterns will be assigned as the features. Furthermore, extracted features will be subjected to a deep learning architecture termed Deep Belief Network (DBN) [23], which predicts the type of industries a location suits. As a main contribution, the weight of DBN will be optimized by a renowned optimization algorithm called Modified Flower Pollination Algorithm (FPA), so that the classification accuracy will be maximum. FPA [24] is a meta-heuristic algorithm inspired by the pollination process of flowers. In order to maximize prediction accuracy, the error between the predicted and actual outcome will be intended to minimize, which is considered the main objective function of this current research.

10.5 EXPECTED OUTCOME

The proposed prediction model will be carried out in the JAVA programming platform, and investigation will be carried out. In the analysis part, the performance of the proposed model will be compared over the existing models in terms of Type I and Type II measures.

REFERENCES

[1] Florian Hillen, Bernhard Höfle, "Geo-reCAPTCHA: Crowdsourcing Large Amounts of Geographic Information from Earth Observation Data", *International Journal of Applied Earth Observation and Geoinformation*, vol.40, pp.29–38, August 2015.

[2] Carsten Kebler, Carson J. Q. Farmer, "Querying and integrating spatial–temporal information on the Web of Data via time geography", *Web Semantics: Science, Services and Agents on the World Wide Web*, vol.35, Part 1, pp. 25–34, December 2015.
[3] Jizhe Xia, Chaowei Yang, Qingquan Li, "Using spatiotemporal patterns to optimize Earth Observation Big Data access: Novel approaches of indexing, service modeling and cloud computing", *Environment and Urban Systems*, August 2018.
[4] Xingjian Lu, Dongxu Jiang, Gaoqi He, Huiqun Yu, "GreenBDT: Renewable-aware scheduling of bulk data transfers for geo-distributed sustainable datacenters Sustainable Computing", *Informatics and Systems*, July 2018.
[5] Rachsuda Jiamthapthaksin, Christoph F. Eick and Seungchan Lee, "GAC-GEO: A Generic Agglomerative Clustering Framework for Geo-Referenced Datasets", *Knowledge and Information Systems*, vol.29, no.3, pp.597–628, December 2011.
[6] Lavika Goel, Mallikarjun Swamy and Raghav Mantri, "Swarm and Artificial Immune System-Based Intelligence Techniques for Geo-Spatial Feature Extraction", *Proceedings of International Conference on Computational Intelligence and Data Engineering*, pp 65–84, December 2017.
[7] Xuan Shi, "Parallelizing Affinity Propagation Using Graphics Processing Units for Spatial Cluster Analysis over Big Geospatial Data", *Advances in Geocomputation*, pp 355–369, January 2017.
[8] Rabindra K. Barik, Ankita Tripathi, Harishchandra Dubey, Rakesh K. Lenka. Tanjappa Pratik and Suraj Sharma, "MistGIS: Optimizing Geospatial Data Analysis Using Mist Computing", *Progress in Computing, Analytics and Networking*, pp 733–742, April 2018.
[9] Songnian Li, Suzana Dragicevic, Francesc Antón Castro, Monika Sester, Tao Cheng, "Geospatial Big Data Handling Theory and Methods: A Review and Research Challenges ISPRS", *Journal of Photogrammetry and Remote Sensing*, vol.115, pp.119–133, May 2016.
[10] Hua-Dong Guo, Li Zhang, Lan-Wei Zhu, "Earth observation big data for climate change research Advances in Climate Change Research, vol.6, no.2, 108–117, June 2015.
[11] Gyozo Jordanm, and JRC PECOMINES Project, "Sustainable Mineral Resources Management: From Regional Mineral Resources Exploration to Spatial Contamination Risk Assessment of Mining", *Environmental Geology*, July 2009.
[12] Mudabber Ashfaq, Ali Tahir, Faisal Moeen Orakzai, Gavin McArdle and Michela Bertolott, "Using T-Drive and BerlinMod in Parallel SECONDO for Performance Evaluation of Geospatial Big Data Processing", *Spatial Data Handling in Big Data Era*, pp 3–19, May 2017.
[13] Jae-Gil Lee, Minseo Kang, "Geospatial Big Data: Challenges and Opportunities", *Big Data Research*, vol.2, no.2, pp.78–81, June 2015.
[14] Mazin Alkathiri, Jhummarwala Abdul and M.B. Potdar, "Geo-spatial Big Data Mining Techniques", *International Journal of Computer Applications*, vol.135, no.11, February 2016.
[15] H. Sun, X. Sun, H. Wang, Y. Li and X. Li, "Automatic Target Detection in High-Resolution Remote Sensing Images Using Spatial Sparse Coding Bag-of-Words Model," in *IEEE Geoscience and Remote Sensing Letters*, vol. 9, no. 1, pp. 109–113, Jan. 2012.
[16] Y. Li, X. Sun, H. Wang, H. Sun and X. Li, "Automatic Target Detection in High-Resolution Remote Sensing Images Using a Contour-Based Spatial Model," in *IEEE Geoscience and Remote Sensing Letters*, vol. 9, no. 5, pp. 886–890, Sept. 2012.
[17] L. Wan, T. Zhang, Y. Xiang and H. You, "A Robust Fuzzy C-Means Algorithm Based on Bayesian Nonlocal Spatial Information for SAR Image Segmentation," in *IEEE Journal*

of Selected Topics in Applied Earth Observations and Remote Sensing, vol. 11, no. 3, pp. 896–906, March 2018.

[18] K. Chen, K. Fu, M. Yan, X. Gao, X. Sun and X. Wei, "Semantic Segmentation of Aerial Images With Shuffling Convolutional Neural Networks," in *IEEE Geoscience and Remote Sensing Letters*, vol. 15, no. 2, pp. 173–177, Feb. 2018.

[19] S. Chen *et al.*, "Interactive Visual Discovering of Movement Patterns from Sparsely Sampled Geo-tagged Social Media Data," in *IEEE Transactions on Visualization and Computer Graphics*, vol. 22, no. 1, pp. 270–279, 31 Jan. 2016.

[20] S. Zhang, J. Li, H. Li, C. Deng and A. Plaza, "Spectral–Spatial Weighted Sparse Regression for Hyperspectral Image Unmixing," in *IEEE Transactions on Geoscience and Remote Sensing*, vol. 56, no. 6, pp. 3265–3276, June 2018.

[21] Z. Li, Z. Liu and W. Shi, "A Fast Level Set Algorithm for Building Roof Recognition From High Spatial Resolution Panchromatic Images," in *IEEE Geoscience and Remote Sensing Letters*, vol. 11, no. 4, pp. 743–747, April 2014.

[22] J. Li, M. Khodadadzadeh, A. Plaza, X. Jia and J. M. Bioucas-Dias, "A Discontinuity Preserving Relaxation Scheme for Spectral–Spatial Hyperspectral Image Classification," in *IEEE Journal of Selected Topics in Applied Earth Observations and Remote Sensing*, vol. 9, no. 2, pp. 625–639, Feb. 2016.

[23] Yara Rizk, Nadine Hajj, Nicholas Mitri, Mariette Awad, "Deep Belief Networks And Cortical Algorithms: A Comparative Study for Supervised Classification", *Applied Computing and Informatics*, March 2018.

[24] A. Y. Abdelaziz, E. S. Ali, S. M. Abd Elazim, "Flower Pollination Algorithm to Solve Combined Economic and Emission Dispatch Problems Engineering Science And Technology, an *International Journal*, vol.19, no.2, pp. 980–990, June 2016.

11 Starring Role of Internet of Things (IoT) in the Field of Biomedical Peregrination for Modern Society

[1]Lipsa Das, [2]Sushree Bibhuprada B. Priyadarshini, [3]Brojo Kishore Mishra, [4]Mahusmita Sahu, and [5]Aradhana Behura

[1]Computer Science and Engineering, Siksha 'O' Anusandhan (Deemed to be University), Bhubaneswar, India

[2]Computer Science and Information Technology, Siksha 'O' Anusandhan (Deemed to be University), Bhubaneswar, India

[3]Computer Science and Engineering, GIET University, Gunupur, India

[4]Technology, Siksha 'O' Anusandhan (Deemed to be University), Bhubaneswar, India

[5]Computer Science and Engineering, Veer Surendra Sai University of Technology, Burla India

CONTENTS

11.1 INTRODUCTION

This section presents the rising future of the Web, known as the "Web of Things", which can associate everything and everybody. IoT inserts insight in the sensor gadgets for conveying data, trading it and taking smart choices. IoT advances human-human correspondence, human-biomedical correspondence and gives excellent association among patients as well as proficient specialists. The IoT requires huge research endeavors in order to handle difficulties. Furthermore, it can give critical individual, expert and financial advantages sooner rather than later. The expression "Web of Things" was initiated by Kevin Ashton of Procter and Gamble in 1999, although he captured the articulation as the "Internet for Things".

The IoT makes the development of web devices, such as PCs, tablets and smart phones to any extent of regular physical contraptions. Counting with development, and the devices that can talk with each other and can be controlled and monitored. Further, IoT has made various advances like examination and statistics, machine learning and embedded systems incorporating various fields of real life. In this context, smart correspondence systems are also required for patients in a down-to-earth way.

So that we can be prepared to give our disease issue to masters which is particularly needed for our overall population. Furthermore, a smart correspondence structure helps us to lessen time just as to diminish space as well. By using IoT we can compose new smart contamination checking and progress of innate structuring and mechanical innovation in like manner to tranquilize the technology in the right way.

In such junction, thinking "Web of Things" as a reality in time whence various contraptions were related to the internet than human is of great interest. It is convinced that the IoT was "conceived" as some place in the scope of 2008 and 2009. Recently, various created and composed approaches inside the degree of the digital change pertaining to healthcare are starting to be used regarding data points of view where IoT accepts a growing employment. Moreover, in expressing applications, such as smart pills, sharp room care, singular human administrations, mechanical innovation and Consistent Prosperity Systems (RTHS), there exist many instances of IoT in social insurance that show what prescription is getting to be fit from gratitude to innovation.

11.2 APPLICATIONS OF IOT IN MEDICAL DOMAIN

11.2.1 Treating Cancer

In July 2018, information was introduced at the ASCO annual gathering from an arbitrary medical investigation of 357 ill people accepting care for head as well as neck malignant growth. The investigation employed a bluetooth-enabled weight scale as well as circulatory strain sleeve, along with an indication voting app, to transmit information to concerned doctors on side effects conjointly with side effects to treatment every weekday.

The patients making use of such smart architecture, known as Cycore, feel less extreme side effects with both the severe rise and the concerned treatment while being contrasted with a control surrounding of ill individuals those move with customary week by week doctor arrivals. Bruce E. Johnson, Leader of ASCO (the American Culture of Clinical Oncology) rearranged care for the two patients as well as their consideration suppliers through rising reactions to be distinguished and aimed to genuinely carry out the weight of such treatment.

The research shows the significant pros pertaining to smart invention discovery for escalating patient contact with physicians while speculating patients' states, which leads to insignificant obstruction with their day-to-day lives. Since Richard Cooper told regarding e-consultancy in a meeting about the future of well-being tech. A portion of the improvements marked get halted persons being attached to their home, or shielded them from being normal in hospital. They are explaining the things that are now and again of very fundamental concerns. Technology forms the association with the therapeutic person to be significantly more important and useful.

11.2.2 Smart Ceaseless Glucose Checking (CGM) and Insulin Pens

Diabetics can be demonstrated as a ripe basis aimed at improvement of relevant appliances as a state which impacts one out of ten individuals, and the person which needs nonstop checking and arrangement of being served. A Consistent Glucose Screen (CGM) represents a device which causes diabetics to constantly screen the blood glucose balance for a few days. The first CGM architecture was endorsed by the US Nourishment and Medication Organization (FDA) in 1999, and as of late, several smart CGMs have reached the market. Smart CGMs such as Eversense as well as Free-form Libre report data on blood glucose levels to any application through iPhone, Android and Apple Watch, enabling the wearer to easily track concerned data as well as differentiate patterns.

The Free-form Libre Link appliance additionally considers remote tracking through guardians, that can include the guardians of diabetic youngsters or the belongings of older patients. Such gadgets are not sustaining the initiation to end up accessibility of the NHS: on World Diabetes Day 2018 (14 November), the NHS can make the Free-form Libre smart CGM accessible on remedy to Type 1 Diabetes sufferers.

It is assessed that this can expand the level of diabetes patients those consider smart CGM gadgets in Britain beginning from 3–5% to 20–25%. Another smart gadget as of now for enriching the lives of diabetes patients is the smart insulin pen. Smart insulin pens—or pen tops—such as: Gocap, InPen and Esysta can naturally record the time, sum and kind of insulin infused in a portion, and prescribe the right sort of insulin infusion at the opportune moment. The gadgets connect with a cell phone application which can store long-haul information while assisting diabetic patients to ascertain their insulin portion, and even (on account of the Gocap) that enable patients to record their suppers and glucose levels, to perceive the way their sustenance and insulin admission impact on the glucose level.

11.2.3 Closed-Circle Insulin Conveyance

One of the crucial territories in IoT prescription involves open-source activity e.g. Open APS, that signifies Open Fake Pancreas Framework. Open APS represents a kind of shut circle insulin transmitting architecture, that differs from a CGM in estimating glucose in a patient's whole circulation framework. Further, it transmits insulin—accordingly "shutting the circle". Open APS got started in 2015 by Dana Lewis and her partner Scott Leibrand, who hacked Dana's CGM and her insulin siphon for robotizing the conveyance of insulin into her structure.

Employing the data collected through CGM as well as Raspberry Pi PC, the own product stops the circle and persistently adjusts the amount of insulin that Dana's siphon transmits. Mechanization of insulin transmittal affords several advantages that can alter the lives of diabetics. Through observing a human's blood glucose levels and naturally altering the amount of insulin transferred into their framework, the APS stores blood glucose inside a sheltered range, anticipating highs and lows (also known as hyperglycemia—undesired high glucose—and hypoglycemia—too much low glucose).

The programmed conveyance of insulin likewise makes diabetics stay asleep from sundown to sunset without the threat of their glucose fall. In spite of the fact that Open APS is not an "out of the container" settlement, it still anticipates individuals to wanting to assemble their very own architecture. It is considering a progressing framework of diabetics who are utilizing it as free and open-source innovation to hack concerned insulin transmittal.

The OpenAPS website states that, "As of January 15, 2018, there are more than (n=1)*1,078+ people in the world over with different kinds of DIY shut circle usage." The OpenAPS group persons are not the major ones to have had such thought. Likewise, in 2013, Bryan Mazlish, a father with a spouse and young child who both have Type 1 Diabetes, formed the major computerized as well as cloud-associated shut circle fake pancreas gadget. In 2014, he established Smart Loop Labs—now known as Bigfoot Biomedical—to scale and market the peregrination of a robotized insulin conveyance framework dependent on his innovation. The organization at present plans for an essential preliminary of its response, subtleties of which are due to be announced in "late 2018 or mid 2019".

11.2.4 Associated Inhalers

The Propeller Wellbeing is one of the crucial makers of smart inhaler innovation. As opposed to delivering entire inhalers, Propeller has designed a sensor which joins to an inhaler or bluetooth spirometer. This interfaces up to an application and assists people with asthma. The organization was set up in 2010, and in 2014 received FDA freedom for two sensors needed to work with inhalers from relevant pharma organizations namely: GlaxoSmithKline's Diskus inhaler, as well as the Respimat inhaler from Boehringer Ingelheim.

Furthermore, Propeller has continued to team up with several makers of inhalers, and now says that its sensor that "works with most inhalers and driving bluetooth spirometers". One of the pros of employing an associated inhaler is enhanced adherence—as it were. Here medicine is taken more reliably and frequently. The Propeller sensor generates an account of inhaler that can be imparted to a patient's primary care physician, and show whether they are employing it as frequently as is endorsed. It gives inspiration to the ill person's lucidity and indicates how the utilization of their inhaler is straightforwardly enriching their state.

11.2.5 Ingestible Sensors

Proteus Advanced Wellbeing and its ingestible sensors represent another scenario of how smart medication can screen adherence. Agreeing to examine through the World Wellbeing Association in 2003, half of prescriptions were not taken as desired. Proteus' framework represents one exertion to diminish such figures. The organization has designed pills which break down in the stomach and generate a little sign which can be received through a sensor worn on the body. Afterwards, the information is passed to a cell phone application, confirming that the patient has accepted the prescription as coordinated.

Proteus has so far attempted such structure associated with pills for treating hypertension that is uncontrolled, Type 2 Diabetes, and antipsychotic medicine. In late 2017, ABILIFY MYCITE—an antipsychotic drug manufactured by Proteus and Otsuka Pharmaceutical Co. became the principal FDA-confirmed sedative with a computerized structure. Similarly, with concerned inhalers, ingestible sensors can follow and enhance how generally patients take their drug, just as enabling them to have an escalating educated discourse with the doctor regarding treatment. While ingesting pills with a sensor may seem to be intrusive, the architecture is chosen with respect to patients, and it can cease sharing a few kinds of data, or stop the program, during associated contact focal points.

Therapeutic smart contact focal points represent a good usage of the Web of Things in a medicinal services setting. Although the notion has great potential, still the science has not decided how to gratify such wishes. In 2014, Google Life Sciences (currently called Verily, a backup of Google's parent organization Letters in order) announced that it would frame a smart contact focal point that can quantify tear glucose and afford an early cautioning architecture to diabetics for alarming them during the time their blood glucose levels are falling or ascending or passed a particular limit. Further, it banded together with Alcon, the eyecare division of pharmaceutical organization Novartis, for its working.

In any situation, the task pulled in huge number of doubts from physicians those accepted that calculated blood glucose levels through tears, was not deductively stable. At length, they were inferred correct. After a protracted period with no true news regarding task peregrinations, in November 2018 it is assured that the undertaking was being racked. Moreover, first framed in 2010, Triggerfish, which is presently CE-stamped and FDA-endorsed. This reveals that it is approved for advertising and deals in Europe as well as the USA., and was approved available to be purchased in Japan in September 2018.

11.2.6 The Apple Watch Application that Screens Despondency

Wearable innovation does not normally need to be planned considering a restorative use to gain social insurance benefits. Takeda Pharmaceuticals USA. and Discernment Unit Restricted, a stage for estimating intellectual wellbeing, worked together in 2017 to probe into the usage of an Apple Watch application for observing and surveying patients with Real Burdensome Issue (MDD). The result from the exploratory examination were presented in November 2017 at pharma and biotech meeting CNS Summit. The research established an abnormal state of consistence with the application, whose members utilized every day to screen temperament conjointly with perception.

The person's every day appraisals were further found to compare with additional top-to-bottom and target cognizance tests and patient-established outcomes, demonstrating that psychological tests communicated by means of an application can even now be hearty and solid. While the investigation was just an exploratory approach, it has exhibited the potential for wearable tech to be employed to assess the effects of melancholy continuously. Similar to smart restorative gadgets that accumulate data, the Apple App application can provide more understanding to patients and insurance experts while empowering the concerned collaborations.

11.2.7 Coagulation Testing

In 2016, Roche launched a Bluetooth-enabled coagulation framework that prepares patients to check how rapidly the blood coagulates. It represents the primary gadget of its kind for hostile to coagulated patients, with self-testing. It appears to enable patients to remain inside their helping range and diminishing the danger of stroke or dying. Having the option to transfer outcomes to medicinal services suppliers indicates curtailed visits to the center. The gadget similarly enables patients to embed comments to their outputs while reminding them to test.

11.2.8 Apple's Research Kit and Parkinson's Ailment

Apple included another 'Development Issue Programming interface' to its open-source Exploration Unit Programming interface in 2018, that permits Apple Watches to screen Parkinson's Ailment indications. Ordinary side effects get checked through the doctor at a center by means of physical analytic tests, meanwhile the patients are urged to maintain a journal to give a more extensive knowledge into manifestations after some time. The programming interface intends to frame that procedure programmed and consistent.

An application on an associated iPhone can exhibit the information in a diagram, giving day-by-day and hourly breakdowns, just as moment by minute manifestation change. Apple's Research Kit has additionally been utilized in various diversified wellbeing contemplates, including a joint inflammation study completed in organization with GSK. Apple is quick to tout the potential for its applications to assist with restorative research and care, and thus, keeping that in mind, in 2017 it launched Care Kit, which represents an open-source structure intended to assist engineers for creating applications for overseeing ailments.

Not at all like Health Kit, which is pointed more at general wellness and prosperity, Care Kit can be utilized to structure applications with a particular restorative reason—hence, speculating this space for increasingly medicinal advancements which employs iPhone as well as Apple Watch innovation.

11.2.9 ADAMM Asthma Screen

ADAMM represents a wearable smart asthma screen which implies for recognizing the side effects of an asthma assault prior to the beginning, enabling the wearer to oversee it prior to the assault deteriorates. This vibrates to advise the person wearing it regarding an approaching asthma assault, and can transfer an instant message to an assigned carrier in the meantime. Various highlights of the gadget incorporate inhaler location—the gadget is able to distinguish and follow inhaler usage, if the patient cannot recollect whether they have utilized and voice journaling to record things such as alterations, sentiments and practices. This additionally possess an estimation innovation which reveals what 'ordinary' is for the wearer after some gap, enabling it to all the more likely comprehend during the time anything has been altered.

ADAMM operates relevant to any implementation and web-based interface, thus, helping asthma patients with setting drug updates, while visualizing information from the gadget, and helping themselves to remember their treatment plan. The gadget was initially anticipated to accomplish FDA freedom and can be released for shoppers toward the finish of 2017, yet has not yet been cleared, demonstrating that such gadgets can now and then set aside a long effort to come to market even once created. On the contrary, an investigation on tolerant wellbeing checking stages that join IoT gadgets distributed in July 2018 notices that ADAMM is "expected to get FDA freedom soon" [3].

11.3 APPLICATIONS OF IOT IN OTHER DOMAINS

11.3.1 Smart Waste Management

Nowadays, waste management is a vital issue in the medical domain. A smart robot which should be cost effective can be employed for effective usage in cleaning the medical coverage area.

11.3.2 Security Application

A thermostat used for reporting on local weather and energy sector controller is a relevant application. The doorbell ring is linked to the internet for authorization. The home smart locking system is also linked to the internet. Various multipronged IoT devices are devised for use by consumers which includes connecting vehicles, wearable devices related to health and fitness with a smart remote/local monitoring system.

11.3.3 Smart Hospital Building

IoT devices are now a part of every hospital automation, including lighting, air conditioning and home security. The power of IoT appliances can be employed for monitoring and controlling of the electronic and electrical frameworks with extended mechanical systems in various types of hospital and building automation systems (e.g., public and private residential, industrial, institutions, etc.).

11.3.4 Old-Age / Disable Care

One of the main applications of IoT enabled smart hospital is to serve quickly in helping the senior citizen and those with special disabilities.

11.3.5 Medical and Healthcare

The **Internet of Medical Things** represent an extension of the IoT in medical as well as health system for data selection and inquiry for care, monitoring and research. IoT devices are used for emergency notification system and remote health monitoring. These range from heart bit monitoring and blood-pressure monitoring to advanced specialized implantations monitoring system like bone-marrow, pacemakers, etc. Similarly, V smart care device is used for whole body UV therapy. This device is used for curing skin disease. It is very costly and it is not available in all hospitals; however, it is very effective.

11.3.6 Smart Transportation

Digital variable speed-limit sign.

Smart transport helps society by reducing the mortality rate of the patient (e.g. pregnant women) effectively. One of the applications of IoT in transportation is digital variable speed-limit sign board. This application of the IoT can lead each aspect of transportation systems. The IoT merges in manufacturing smart medical devices that are equipped with networking as well as communication capabilities for sensing, identification, actuation, control and processing. Working on the advanced smart cyber-physical space, it enables market chances and new business manufacturing units. IoT can be used to design the architecture of the cyber-physical aided manufacturing system.

11.3.7 Smart Agriculture for Medicinal Plant

IoT is used in agricultural farming for data collection on rainfall, temperature, humidity, pest measure and control, wind speed, and soil fertility index. This data is utilized in farming techniques for proper decision making, risk assessment and wastage management and simultaneously; it will reduce the effort required for managing the productivity of crops. IoT enabled farmers can easily monitor the soil index like temperature and moisture from anywhere, and is able to put IoT-collected data for precision fertilization and production. Medicinal plants are very important for medicine production.

For example, in August 2018, Toyota Tsusho started to create a fish farming tool with the partnership of Microsoft by using Microsoft Azure application suite for water management in IoT technologies. Researchers at Kindai University developed a mechanism in a water pump system to count the number of fish in the conveyor belt and apply the artificial intelligence to analyze the count of fish, and calculate the effectiveness of water flow through programming and the data is provided by the fish. The specific computer programs employed in that process, which comes under the Azure Machine Learning platform as well as the Azure IoT Hub platform. This example presents a real life situation.

11.3.8 Smart Infrastructure Deployment

Check and control of operations of various sustainable infrastructures and buildings like tracks, steps, etc is a marvel of the IoT. The IoT can be used for checking any event that compromises safety and increases risk. There are many planned large scale projects of IoT to make better city control. Like the city of Songdo in South Korea, which is a fully connected, wired and highly equipped smart-city. As of June 2018 roughly 70 percent of its business districts are completely inter-connected. Major city is expected to be wired very soon and fully automated with little or no human intervention. A large number of electrical devices already merged with internet connectivity, that facilitates them to collaborate with various utilities to enable power production conjointly with energy distribution as well as energy optimization as a whole.

11.3.9 Environment Monitoring

IoT uses its sensor for measuring, recording and monitoring the environmental factors in environmental conservation system by checking soil and atmospheric conditions as well as the quality of air and water. It also checks and monitors the wildlife habitations and movements.

11.3.10 Living Lab

Another application of merging IoT is 'Living Lab in hospital', that fuses research process and program and setting a public-private-people and doctor partnership. Presently There are 320 'Living Labs' which uses IoT for collaboration and sharing of knowledge among stakeholders to enhance technological and innovative products.

11.4 CONCLUSIONS

We conclude a top-level view of the arrival, software and results of Internet of Things in recent times' fast-paced ever-changing dynamic biomedical location. We need to be alert in the direction of the security and occasion fee of the devices due to the reality records manipulation reasons and severe hassle for our fitness. Thus, we analyze the various ways in which IoT has modified our biomedical vicinity and life and made our work a very excellent deal with much less hard, greater available, more responsible and new manner. Further, we in particular focus on the unfavorable outcomes of IoT inside the healthcare area with its numerous software and groundbreaking outcomes inside the fields of bioinformatics, neural engineering, neurorobotics and so on.

The applicability of the IoT in healthcare domain (the industry, personal healthcare and healthcare rate programs) is increasingly prolonged all through several particular internet of factors use cases at the identical time. This chapter discusses different healthcare IoT instances employed for deciding to speed up and accelerate day-to-day life, despite the fact that hurdles continue to be accelerating. Hence a long manner, maximum IoT tasks in healthcare moved around the development of care as such with remote tracking and tele-monitoring as the primary packages.

REFERENCES

[1] M. Hassanalieragh, A. Page, T. Soyata, G. Sharma, M. Aktas, G. Mateos, B. Kantarci, and S. Andreescu 2015, "Health Monitoring and Management Using Internet of Things(IoT) Sensing with Cloud Based Processing: Opportunities and Challenges," IEEE *Xplore*, pp. 285291.

[2] T. Kriplean et al., "Physical Access Control for Captured RFID Data," *IEEE Pervasive Computing*, vol. 6, no. 4, pp. 48–55, 2007.

[3] econsultancy.com/internet-of-things-healthcare/

[4] K. Chen, K. Nahrstedt, and N. Vaidya. The Utility of Explicit Rate-based Flow Control in Mobile Ad Hoc Networks. In *WCNC '04: Proceedings of the IEEE Wireless Communications and Networking Conference*, volume 3, pages 1921–1926, Mar. 2004.

[5] L. Torrey and J. Shavlik, *Handbook of Research on Machine Learning Applications and Trends*. Hershey, PA: IGI Global, 2010.

[6] Q. Zhang,·L. Cheng, and R. Boutaba, "Cloud computing: state-of-the-art and research challenges," *J Internet Serv Appl*, vol. 1, pp. 7–18, 2010.

[7] https://en.wikipedia.org/wiki/Bioinformatics

[8] https://en.wikipedia.org/wiki/DNA_sequencing

[9] https://en.wikipedia.org/wiki/Neural_engineering

[10] https://en.wikipedia.org/wiki/Genetic_engineering

[11] J. Zheng, D. Simplot-Ryl, C. Bisdikian, and H. Mouftah, "The Internet of Things," in *IEEE Communications Magazine, Volume:49, Issue: 11, pp:*30–31, 2011.

[12] Y. Huang and G. Li, "Descriptive Models for Internet of Things," in *IEEE International Conference on Intelligent Control and Information Processing (ICICIP)*, August 2010.

[13] T. Fan and Y. Chen, "A Scheme of Data Management in the Internet of Things," in *2nd IEEE International Conference on Network Infrastructure and Digital Content*, Sept. 2010.

[14] Y. Huang and G. Li, “A Semantic Analysis for Internet of Things,” in *International Conference on Intelligent Computation Technology and Automation (ICICTA)*, May 2010.

[15] Q. Zhou and J. Zhang, “Research Prospect of Internet of Things Geography,” in *19th International Conference on Geoinformatics*, June 2011.

[16] J. Li, Z. Huang, and X. Wang, “Countermeasure Research about Developing Internet of Things Economy,” in *International Conference on E-Business and E -Government (ICEE)*, May 2011.

[17] Y. Yu, J. Wang, and G. Zhou, “The Exploration in the Education of Professionals in Applied Internet of Things Engineering,” in *4th International Conference on Distance Learning and Education (ICDLE)*, October 2010.

[18] L. Coetzee and J. Eksteen, “The Internet of Things: Promise for the Future? An Introduction,” in *IST-Africa Conference Proceedings, CSIR, Pretoria, South Africa*, May 2011.

[19] L. Tan and N. Wang, “Future Internet: The Internet of Things,” in *3rd International Conference on Advanced Computer Theory and Engineering (ICACTE)*, August 2010.

[20] G. Gang, L. Zeyong, and J. Jun, “Internet of Things Security Analysis,” in *International Conference on Internet Technology and Applications (iTAP)*, August 2011.

[21] M. Wu, T. Lu, F. Ling, J. Sun, and H. Du, “Research on the Architecture of Internet of Things,” in *3rd International Conference on Advanced Computer Theory and Engineering (ICACTE)*, August 2010.

[22] Z. Hu, “The research of several key question of Internet of Things,” in *International Conference on Intelligence Science and Information Engineering (ISIE)*, August 2011.

[23] J. Sosa, H. Bowman, J. Tielsch, N. Powe, T. Gordon and R. Udelsman 1998, “The Importance of Surgeon Experience for Clinical and Economic Outcomes From Thyroidectomy,” *Annals of Surgery*, Volume 228, Number 3, pp. 320330.

[24] W. Sung and Y. Chiang, “Improved Particle Swarm Optimization Algorithm for Android Medical Care IOT using Modified Parameters,” *Journal of Medical Systems*, Volume 36, Issue 6, pp. 3755–3763, 2012.

[25] www.researchgate.net/publication/267943995_Configuration_Management_Process_Improvement_for_the_Medical_Device_Industry.

[26] Rajkumar Buyya, Christian Vecchiola, S.Thamaraiselvi: “Mastering Cloud Computing”, McGraw Hill.

[27] S. Kim, and S. Kim. “User preference for an IoT healthcare application for lifestyle disease management”, *Telecommunications Policy*, 2017.

[1?] [illegible] and [illegible], "A Semantic [illegible] for [illegible] of Things," in International Conference on [illegible] Communication, [illegible] and Applications, [illegible], May 2010.
[15] Q. Zhou and J. Zhang, "[illegible] of Internet of Things [illegible]," in [illegible] International Conference on [illegible], June 2011.
[16] [illegible] and X. [illegible], "Comprehensive Research about Developing Internet of Things [illegible]," [illegible], 2011.
[17] [illegible], [illegible] Wang, and G. Zhou, "The [illegible] of the [illegible] of [illegible] in Applications of Internet of Things [illegible]," in International Conference on [illegible] Computing and [illegible] (ICICTA), October 2011.
[18] [illegible] and [illegible], "The Internet of Things: [illegible] for the [illegible]," [illegible], May 2015.
[19] [illegible] and [illegible] Wang, "[illegible]: The Internet of Things," in [illegible] International Conference on [illegible] (ICACTE), [illegible] 2010.
[20] [illegible], "[illegible] of Things [illegible]," [illegible], 2011.
[21] [illegible], [illegible], and [illegible], "[illegible] Internet of Things," in International Conference on Advanced Computer Theory and Engineering (ICACTE), [illegible] 2010.
[22] X. [illegible], "[illegible] of Internet of Things," in [illegible] Conference on [illegible] (ICI), August 2011.
[23] [illegible], [illegible], [illegible], Magazine, Volume [illegible], [illegible].
[24] [illegible] for [illegible] of [illegible] in [illegible] Technology, [illegible] 2012.
[25] [illegible] "[illegible] Configuration Management [illegible] for the [illegible] Device [illegible]."
[26] [illegible], "[illegible] Management [illegible]," March 2011.
[27] [illegible], "[illegible] in [illegible] of [illegible]," [illegible], 2011.

Index

Note: Page numbers in *italics* indicate figures and in **bold** indicate tables on the corresponding pages.